Making "Sense" of GENETICS TODAY

Karl Chai

innovativeink
PUBLISHING
A Division of Kendall Hunt

innovativeink PUBLISHING
A Division of Kendall Hunt

www.innovativeinkpublishing.com
Send all inquiries to:
4050 Westmark Drive
Dubuque, IA 52004-1840

Library of Congress Control Number: 2024917566
Print ISBN: 979-8-3851-1141-1
Ebook ISBN: 979-8-3851-1142-8

Published in the United States of America

Contents

Chapter 2. How Special Is Human Genetics? 15

Chapter 4. RNAs Are The Sense Molecules of Life 57

Chapter 5. Genetic Variants Affect Protein Structure and Function 73

Preface

On an early December Wednesday afternoon in the Harp & Celt Irish Pub in downtown Orlando where the official Orlando Supporters Club for the Liverpool Football Club meet on game days, I took the opportunity of the team winning the game against a rather weak opponent to ask a few club members the question: when you hear or see the term "genetics," what comes to your mind? I announced to the small crowd that I am writing a book on human genetics to bridge the gap between the cutting-edge advances and the practical understanding by the public. I requested that I would prefer for the responder to not have taken a college class in genetics. I don't know if it helped or hurt my efforts when Chris the goalkeeper (who always wears a goalkeeper's jersey) followed my announcement with a bunch of big words to sell my stature (if I had any) in the field of genetics!

Gerald: "If I am being completely honest, it's 'Jurassic Park.' It was the first time I was introduced to it (genetics). But as I have gotten older, from what I understand, not as an expert, it's about humans, not really anatomy or biology but the DNA side of it, which makes no sense to me. I am sure it makes a lot of sense to some people."

Uno: "What is passed down from my parents and grandparents, etc." (I later learned that Uno just had a new baby son this year—Congratulations, Uno! He also commented on the human struggles against cancer, which his father is currently experiencing.)

Tom: "The first thing that comes to my mind is, you think of 'hereditary,' your parents, my genetics (in such contexts). What really comes to my mind is, I think of Chernobyl. All these creatures that are surviving in that land no one is allowed to survive in, no one. They now have wolves, dogs, and birds, and they are actually surviving. How are they able to survive that? Maybe we can figure out what gene(s) they have that is preventing them from getting cancer, and cure cancer or stop cancers from starting . . ."

Thank you, mates, these are some brilliant responses! I will most likely disappoint some readers by having quite a lot of technical words in the text, but I made every effort at minimizing the technical details in especially the modern biotechnology when it is applied to answering the human genetic questions of today.

I sincerely hope, at the end, the story lines that I put together can help make "sense" of human genetics when we think of the topic today.

Acknowledgments

I wish to thank Angela Lampe, Director of Sales and Strategic Partnerships at Innovative Ink Publishing for inspiring me to take on this project. And I thank Lynne Rogers for her editorial assistance in the drafting of the manuscript.

1 What Is Genetics?

Garden peas and fruit flies

1.1 Genetics Is Heredity AND Variations

There are quite a number of ways to define genetics, such as:

- National Institute of General Medical Sciences (NIGMS): Genetics is "the scientific study of genes and heredity—of how certain qualities or traits are passed from parents to offspring as a result of changes [underscore added by the author] in DNA sequence."
- National Human Genome Research Institute (NHGRI): "Genetics is the branch of biology concerned with the study of inheritance, including the interplay of genes, DNA variation [underscore added by the author] and their interactions with environmental factors."
- Centers for Disease Control and Prevention (CDC): Genetics is "the study of how genes or groups of genes are involved in health and disease" [underscore added by the author].
- *Genetics for Dummies*: "*Genetics* is the field of science that examines how traits are passed from one generation to the next. Simply put, genetics affects *everything* about *every* living thing on earth. An organism's *genes* are segments of DNA (deoxyribonucleic acid) that are fundamental units of heredity. Genes play an essential role in how the organism looks,

> behaves, and reproduces [underscore added by the author]. Because all biology depends on genes, understanding genetics as a foundation for all other life sciences, including agriculture and medicine, is critical."

Now we have the definitions of "genetics" from two national institutes of health and the CDC and a book titled *Genetics for Dummies.* These definitions are similar and fairly accurate, although the NIGMS definition could use a little tweak. In addition, considering the prestige of the first three sources and the title of the fourth source, it is not unreasonable to say that these probably covered the whole spectrum of all variations of the definition of genetics.

The definitions by the NIGMS and the NHGRI clearly focused on "heredity" and "inheritance," whereas the CDC, as they should, focused on "health and disease." For "Dummies," it is also "how traits are passed from one generation to the next," that is, heredity or inheritance. It is a safe bet that a majority of us would think of heredity, inheritance, or how traits are passed from one generation to the next when asked about genetics. It is also clear that it makes little difference if the surveyed population has taken a college or even high school course in genetics.

Why have I then underscored "changes," "variation," "health and disease," and "looks, behaves, and reproduces" in these various definitions of certain levels of authority (some quite high)? The truth is that genetics as a science is about looking at the differences and changes among the individual representatives of an organism. Heredity and inheritance are not less important, but the concept becomes clear when we go to an extreme to look at this. Going to extremes is a very powerful practice in any branch of science, as it peels away the confounding factors. The extreme scenario is this: Imagine that every human being of the same sex looks exactly the same, would anyone still use the term "genetics"?

The term *genetics* was coined by the English biologist William Bateson (1861–1926) in a personal letter to the zoologist Adam Sedgwick on naming a professorship when Bateson was consulted: "If the Quick fund were used for the foundation of a professorship relating to Heredity and Variations the best title would, I think, be the 'Quick Professorship of the Study of Heredity'. No single word in common use quite gives this meaning. Such a word is badly wanted, and if it were desirable to coin one, 'genetics' might do." In this letter dated April 18, 1905, the professorship was first considered to be relating to "experimental breeding," which was crossed off and replaced with "heredity and variations." It is clear that the "study of heredity" means "heredity and variations" by the newly coined word "genetics." The first suggested title of "experimental breeding" was very telling. If all breeding experiments generate exactly the same things, then there would be nothing experimental to think of or perform.

The two most frequently used genetic terms, "phenotype" and "genotype," also have the meaning of variations built in. After all, if all are the same (looks and behaviors), there would be no "type" to speak of. The Cambridge Dictionary defines the word type as "a particular group of people or things that share similar characteristics and form a smaller division of a larger set." Well, that again translates to "heredity and variations," that is, sharing similar characteristics and forming a smaller division of a larger set. The NHGRI defines genotype as "a scoring of the type of variant present at a given location (*i.e.*, a locus) in the genome" and phenotype as "an individual's observable traits, such as height, eye color and blood type. A person's phenotype is determined by both their genomic makeup (genotype) and environmental factors." We address the meaning of the second sentence later, but clearly height, eye color, and blood type (!) all show differences among humans. The definition of genotype actually accurately defined another genetic term "locus," a given location in the genome where genetic variants are present. A genome is "the entire set of DNA instructions found in a cell" (according to the NHGRI). The NHGRI definition of locus as "a physical site or location within a genome (such as a gene or another DNA segment of interest), somewhat like a street address" is rather inaccurate or inadequate—it does not factor in the variations. This is especially important in experimental genetics, which tests for observable differences resulting from the genetic differences.

1.2 Mendelian Genetics

The honorary title of the "Father of Genetics" is given to Gregor Mendel. His work that earned him that title, first translated in 1901 from German to English by none other than William Bateson, was titled *Experiments in Plant Hybridization* (*Versuche über Plflanzenhybriden*) (1865). In the introductory remarks, it began with "EXPERIENCE OF ARTIFICIAL FERTILIZATION, such as is effected with ornamental plants in order to obtain new variations in color, has led to the experiments which will here be discussed. The striking regularity with which the same hybrid forms always reappeared whenever fertilization took place between the same species induced further experiments to be undertaken, the object of which was to follow up the developments of the hybrids in their progeny." In the Cambridge Dictionary, hybridization is defined as "the process of producing a plant or animal from two different types of plant or animal." Mendel worked with garden peas, *Pisum sativum*, using "characters" "transmitted unchanged to the hybrids and their progeny."

Mendel's study objective was to observe these variations and deduce the law by which the differentiating characters appear in successive generations. He discarded the characters that did not permit a "sharp and certain separation," characters of a "more or less" nature, and selected seven pairs of differentiating characters. The best illustrated two were the form of the ripe seeds, round or roundish *versus* wrinkled, and the color of the seed albumen (endosperm), pale yellow, bright yellow, and orange colored *versus* a more or less intense green tint. To simplify these two pairs of differentiating characters, as commonly seen in modern genetics books, they are round/wrinkled and yellow/green. The plants Mendel used in his "crosses," or hybridization, "possess constant differentiating characteristics," that is, pure bred or, in later genetic terms, homozygous.

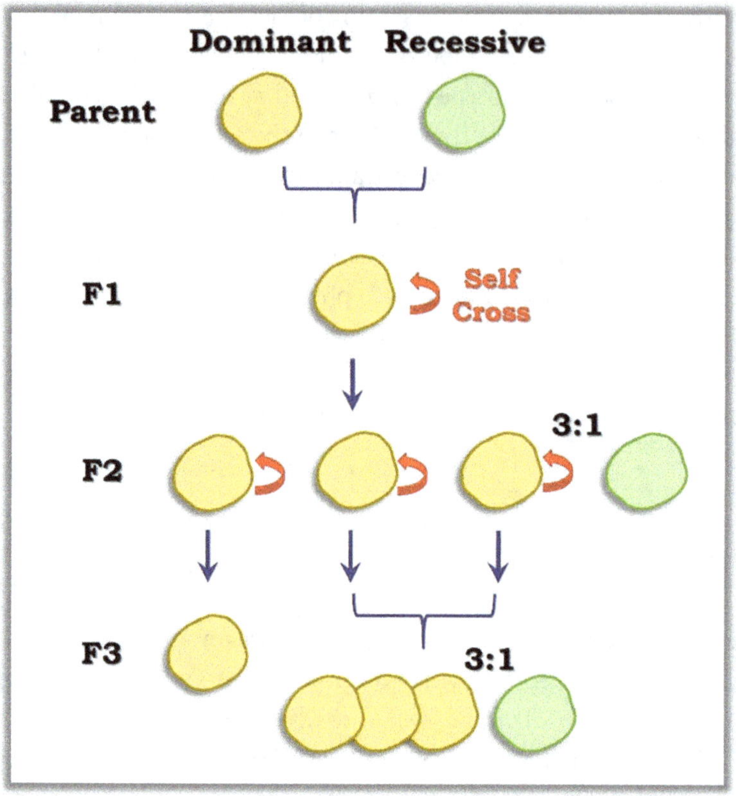

FIGURE 1-1. The escape and reappearance of a recessive character (green) in successive generations of hybrids, as observed by Gregor Mendel

1.2.1 DOMINANT *VERSUS* RECESSIVE TRAITS

Mendel's first principle finding was the characters that he defined as "dominant," such as the round form and the yellow color of the seeds (**Figure 1-1**). These characters are shown in the hybrid (F1), while the respective counterparts, the wrinkled form and the green seed color "escape observation completely," and were defined as "recessive." The choice of the word recessive to describe the characters not shown in the hybrid was very careful and deliberate. These characters "withdraw or entirely disappear in the hybrids, but nevertheless reappear unchanged in their progeny"; here, "their" refers to the hybrids. The dominance or recessiveness was independent of the parental line, the characters shown in the hybrid were identical whether the dominant character was contributed by a seed plant or a pollen plant. Mendel paid tribute to Carl Friedrich von Gärtner (1772–1850, what a perfect name), who pioneered the work on plant hybrids, on his original observation of the sex independence of the dominant characters.

The crosses performed on all seven pairs of differentiating characters produced similar results in the appearance of the dominant character. In the first generation (F2) from the hybrids (F1), among each four plants, three would display the dominant character and one the recessive. In the second generation (F3) from the hybrids (F1), the three dominant portions were shown to be one portion of constant dominant and two portions giving the same 3:1 ratio of dominant *versus* recessive characters as the first generation (F2) from the hybrids (F1). Thus, in the F2 generation, the observed ratio should be interpreted as 1:2:1, most familiar to students of genetics in modern times, and as shown in modern genetics textbooks.

1.2.2 INDEPENDENT SEPARATION

These hybridization experiments provided the information for what is now known as Mendel's First Law of Genetics, the law of (independent) segregation. Actually the translated word used (by Bateson) in Mendel's paper to describe the behaviors of the pairs of differentiating characters was "separation." It simply points to the fact that the dominant and recessive characters of a pair in the hybrid do not mix and that the recessive character can reappear in the progeny of the hybrid. Today in all the genetics books, for college courses or "dummies," the word of choice is segregation. If the principle of genetics ends with the first law, it would make very little difference in which word to use. But the difference of these two choices becomes clear when we continue to (re)examine Mendel's work.

In the next sets of experiments, Mendel "united" several diverse (pairs) of characters in the hybrid to see if the "law of development" (separation) observed with each pair of differentiating factors held true. The first important observation was that uniting the diverse pairs of characters did not affect the separation for each pair, reproducing the same ratios in the generations from the hybrids, in a sex-independent manner.

1.2.3 INDEPENDENT ASSORTMENT

The second and most important finding was, in the translated words of Mendel's, "If several differentiating characters are combined by cross-fertilization in a hybrid, the resulting offspring form the terms of a combination series in which the combination series for each pair of differentiating characters are united." Mendel extrapolated from these observations to state that the "law of development" would also apply to the more intermediate, "less sharply defined" pairs of characters.

The experiments Mendel used to validate the conclusions were ingenious, most remarkable, most remarkably ahead of the times, and immortal (okay, I ran out of adjectival words). The brilliance of Mendel's has, however, long been given a tremendous injustice in modern genetics textbooks. In most if not all of them, a "Second Mendelian Law of Genetics: Independent Assortment" is illustrated with a 16-space Punnett Square showing the results of crossing the hybrids (F1) of two pairs of characters. This is quite confusing and is actually not how informative genetic crosses are performed, by Mendel, or modern geneticists in the pursuit of "disease genes" in humans. Of course, in humans, no test crosses are actually "performed," but the results of natural mating are considered.

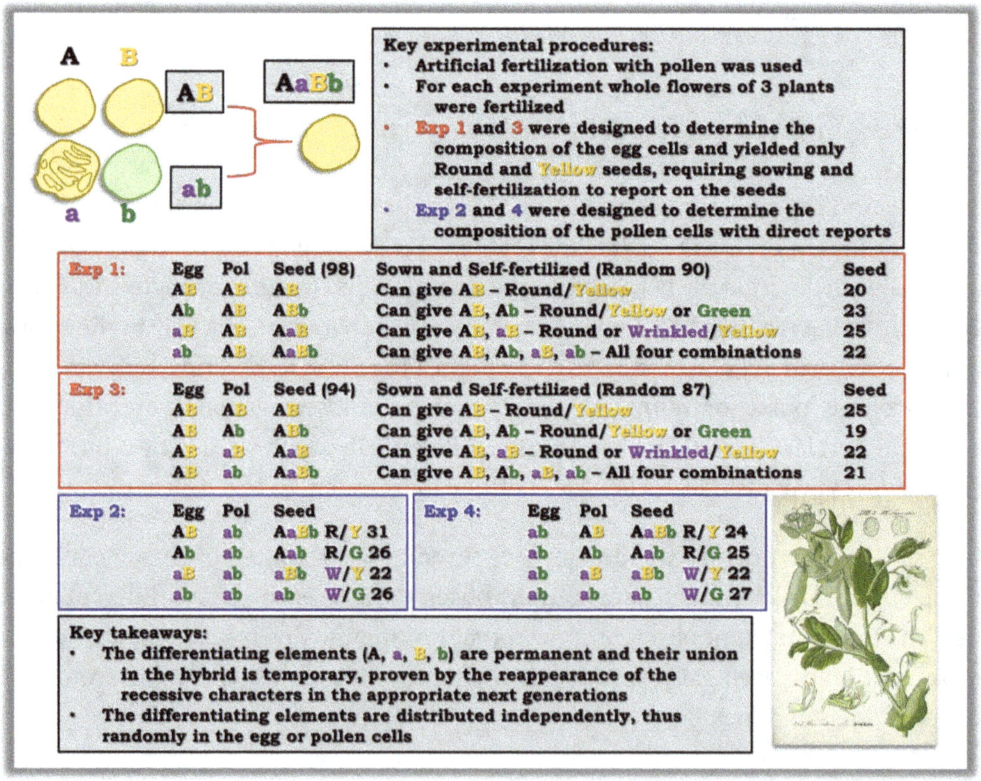

FIGURE 1-2. Mendel's key experiments with *Pisum sativum*

First, Mendel considered the elements rendering the constant characters to be permanent and unaffected by the union of the opposing constant elements in the hybrid. He was the first to use the capital letters to designate the dominant characters (elements) and the small letters for the recessive characters (elements): in his experimental plants, "A" = round and "a" = wrinkled; and "B" = yellow and "b" = green. We must emphasize that Mendel used single letters instead of a pair for the purebreds, as this was decades before the "chromosome hypothesis of heredity" to give us the luxury of knowing that the plants are diploid and that the pollen or the egg is haploid.

The key experiments are outlined in **Figure 1-2**, the initial crosses involved artificial fertilization, that is, pollens from plants with chosen characters were used to fertilize plants with the paired characters. The constant "AB" (round/yellow, R/Y) parent was crossed with the constant "ab" (wrinkled/green, W/G) parent to produce the "AaBb" hybrid. Again, at the time of Mendel's experiments, these paired "Aa" and "aB" letters did not have the meaning of diploidy. For each of the following experiments, whole flowers of three plants were artificially fertilized.

<u>Experiment 1:</u> Crossing the hybrid with the pollen of "AB." This is translated to four different egg cells "AB," "Ab," "aB," and "ab" to combine with one kind of pollen cell "AB," yielding four possible combinations of the differentiating elements, "AB," "ABb," "AbB," and "AaBb." All would produce only R/Y seeds. The number of element combinations in the egg cell was simply the maximum possible, from two sets of characters with two kinds each, that are "permanent and unaffected by the union" in the hybrid. It is the same consideration in the other experiments, as well.

<u>Experiment 2:</u> Crossing the hybrid with the pollen of "ab." This is translated to four different egg cells "AB," "Ab," "aB," and "ab" to combine with one kind of pollen cell "ab," yielding four possible combinations of the differentiating elements, "AaBb," "Aab," "abB," and "ab." All four possible combinations of characters of the two pairs would be represented, namely, R/Y, R/G, W/Y, and W/G.

<u>Experiment 3:</u> Crossing "AB" with the pollen of the hybrid. This is translated to one kind of egg cell "AB" to combine with four different pollen cells "AB," "Ab," "aB," and "ab," yielding four possible combinations of the differentiating elements, "AB," "ABb," "AbB," and "AaBb." All would produce only R/Y seeds.

<u>Experiment 4:</u> Crossing "ab" with the pollen of the hybrid. This is translated to one kind of egg cell "ab" to combine with four different pollen cells "AB," "Ab," "aB," and "ab," yielding four possible combinations of the differentiating elements, "AaBb," "Aab," "abB," and "ab." All four possible combinations of characters of the two pairs would be represented, namely, R/Y, R/G, W/Y, and W/G.

Mendel's intent was to determine the composition of the hybrid egg cells (experiments 1 and 2) and of the hybrid pollen cells (experiments 3 and 4). He was the first in the documented history to experimentally inquire into the hereditary composition of the reproductive cells. For this, Mendel fully earned his title of Father of Genetics. In ancient Greece, Hippocrates (460–375 BCE), known as the father of medicine, believed in pangenesis, that is, all organs of a parent during sexual intercourse pass on some hereditary components, which reassemble in the uterus to form a baby. This hypothesis of pangenesis would suggest that the blood is vitally important in transmitting heredity, which was believed by Aristotle (384–322 BCE). The semen was regarded as purified blood and the menstrual blood was the female counterpart, while the two parts combine in the uterus to form a baby. The influence of such early thoughts of how the blood was the material being passed on to the next generation remains worldwide in today's society, and even the scientific community. The current marriage laws in the Civil Code of the People's Republic of China prohibit the union between "直系血亲," that is, direct blood relatives. In the scientific literature, mating between "blood relatives" is described as "consanguineous," from Latin for "together" (con-) and "blood" (sanguis).

1.3 Chromosomes Carry the Genes—The Chromosome Hypothesis of Heredity

Mendel's work was before the discovery of DNA, which was not too far behind, and before the discovery of the chromosomes, especially the observation of chromosome behaviors and distributions during the formation of the reproductive cells. DNA was discovered in 1869 from the pus on wound bandages by the Swiss physician Friedrich Miescher (1844–1895), who also isolated DNA from fish sperm. But Miescher came into doubt about DNA being the material of heredity, pretty much due to its lifeless look in the purified and dried form. We will have a more in-depth discussion on the DNA in a later chapter.

Chromosomes were discovered by the German cytologist Johann Adam Otto Bütschli (1848–1920) who studied unicellular organisms, such as the ciliates. Bütschli was the first to describe mitosis, the division of cells, and the spindle, equatorial plate, metaphase, anaphase, and telophase in the division. The term *chromosome* came from the German anatomist Heinrich Wilhelm Gottfried von Waldeyer-Hartz (1836–1921) who studied the basophilic stained filaments known as chromatin, described by the German biologist Walther Flemming (1843–1905). The term *chromatin* was retired and then given a new meaning in the modern era of molecular genetics, or, genomics. We will have a more in-depth look at the meaning of this term in a later chapter.

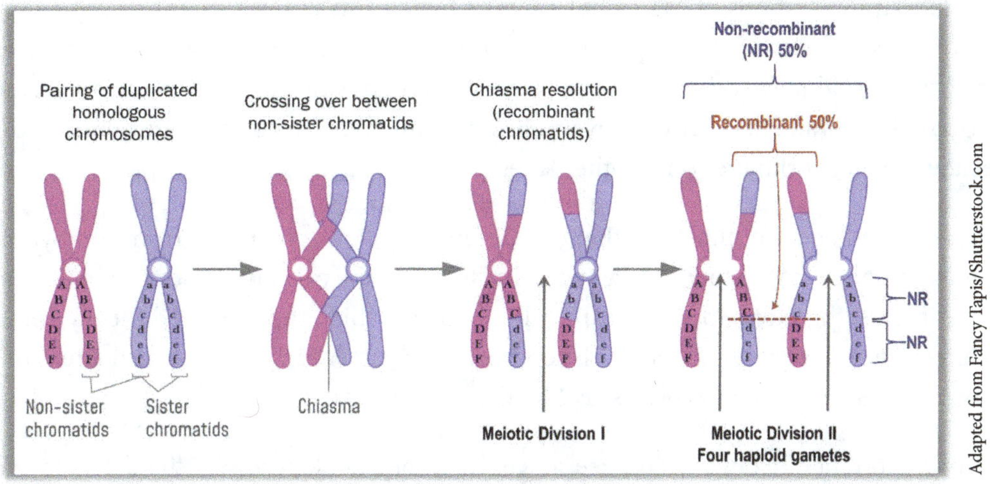

FIGURE 1-3. Chromosome distribution in meiosis, the cell divisions that produce haploid reproductive cells (gametes) in diploid organisms, including garden peas, fruit flies, and humans

1.3.1 SUTTON-BOVERI-MONTGOMERY

The foundation of modern genetics was laid by the American cytogeneticist Walter Stanborough Sutton (1877–1916), who, in a manner of speaking, put two and two together and described the association between the behaviors of the differentiating elements in Mendel's experiments and the chromosome behaviors during the formation of the reproductive cells in the grasshopper *Brachystola magna*. This is known as the Sutton-Boveri Chromosome Hypothesis of Heredity. Sutton was inspired by a lecture in September 1902, given by William Bateson in New York, and gave proper credit to prior observations by the German zoologist Theodor Boveri (1862–1915) and American zoologist Thomas Montgomery (1873–1912). In Sutton's paper, he concluded: "In *Brachystola* no qualitative division of chromosomes but only a separation of the two members of a pair which, while coexisting in a single nucleus, may be regarded as jointly controlling certain restricted portions of the development of the individual."

Today, this is the well-known and well-described process of meiosis, a special type of cell division in the formation of reproductive cells (**Figure 1-3**). During meiosis, each chromosome is duplicated with the DNA therein replicated, and the homologous chromosomes pair up before the first division. The first division pulls apart the paired and duplicated homologous (same numbered) chromosomes, with chromosomal material exchanges in between, a phenomenon observed by Alfons Janssens (1863–1924) a few years later. The duplicated chromosomes then separate in a second division to give the reproductive cells that contain exactly half of the chromosomes as that in the first parent (pregametic) cell.

In the modern illustrations of Mendel's laws of genetics, it is customary to use a pair of letters to denote the parent plants, for example, Yy for yellow/green, and Rr for round/wrinkled, and single letters in the reproductive cells to show their genetic composition. This not only makes sense after the proposed and validated Sutton-Boveri chromosome hypothesis of heredity but also makes us appreciate Mendel's brilliance on a higher level.

Mendel expected equal numbers of the four different egg or pollen cells from the hybrids and in turn, equal numbers of the four different combinations in the progeny. And the results were: "In all the experiments, therefore, there appeared all the forms which the proposed theory demands, and they came in nearly equal numbers." The emphasis was on "nearly," given the degrees of difficulty in cultivating the experimental plants, as Mendel acknowledged.

The most important outcome was the reemergence of the double recessive characters in the crossing of a hybrid, egg or pollen, to the double recessive parent, egg or pollen, all expressed as "ab." That is to say, the elements that give the recessive characters went into the hybrid, surely from the parent, and seemed to have disappeared. But they must not have, as the hybrids, while manifesting all dominant characters, produced egg or pollen cells that could contain only the recessive elements. Otherwise, the back cross of a hybrid with the double recessive parent could not have generated the double recessive progeny. Mendel concluded: "We must further assume that it is only possible for the differentiating elements to liberate themselves from the enforced union when the fertilizing cells are developed. In the formation of these cells all existing elements participate in an entirely free and equal arrangement, by which it is only the differentiating ones which mutually separate themselves." Furthermore, he concluded: "Some justification of the opinion expressed lies in the evidence afforded by *Pisum* that the behavior of each pair of differentiating characters in hybrid union is independent of the other differences between the two original plants, and, further, that the hybrid produces just so many kinds of egg and pollen cells as there are possible constant combination forms."

1.3.2 THE CHROMOSOME LAW OF GENETICS

In summary, the elements that endow the differentiating characters are permanent and thus independent of one another, either in a pair or against other pairs, in the observed behaviors during the formation of the egg or pollen cells. The Law of Genetics at the chromosomal level can be stated as "for chromosomally diploid organisms reproducing via a haploid gametic step it governs the distribution of chromosomes in the gametes, with each gamete receiving one and only one but either one from each pair of homologous chromosomes." The significance of the word segregation in the framing of the law of genetics, a better choice than separation, is the emphasis on each gamete having to receive a complete slate of all chromosomes from every pair.

This book is intended for making "sense" of genetics and not as a thorough review of the history of genetic discoveries. We look at key historical moments in genetics to justify rewriting or restating The Law of Genetics to help make sense of something quite simple and straightforward but somehow made out to be more complex than necessary.

1.3.3 "VIOLATIONS" OF THE LAW OF GENETICS—LINKED LOCI

The Sutton-Boveri chromosome hypothesis of heredity at its inception was imperfect and received strong and sustained criticism. A major opposing view was initially from Thomas Hunt Morgan (1866–1945), as he saw the much smaller number of chromosomes than the pairs of traits (differentiating characters, elements) to be irreconcilable for the model organism in his laboratory, the fruit fly *Drosophila melanogaster*. Unless, many traits from one parental origin "Mendelize together," meaning to segregate, together. Morgan described it as a "fatal objection to the chromosomal theory." A great deal of the benefit of the doubt must be extended to the early pioneers in cytogenetics as the techniques used were far from the highly sensitive and precise means of today. For example, both Sutton and Morgan thought that the separation of the parental chromosomes happens in the second division of meiosis, instead of the first.

Experimental proof came from the Belgian cytologist Frans Alfons Janssens (1863–1924) who observed the "chiasmatype," which is the crossing over of two duplicated parental chromosomes, called the (nonsister) chromatids, with the exchange of parental chromosomal material before the first meiotic division (**Figure 1-3**). This was the key contribution to help reconcile the uni-parental traits being carried on one chromosome, but not "Mendelizing together." Morgan revised his view with reference to the "chiasmatype theory" of Janssens' and anticipated that the "linkage" (togetherness) of the traits depends on the linear difference apart of the chromosome materials that represent them. This was the theoretical basis of "linkage analysis," a method that gave us the Huntington disease and cystic fibrosis genes, decades later.

In the early 1900s, it was applied to produce a genetic map for 30 distinct genes in the four *Drosophila* chromosomes by Morgan and his students and colleagues in the Fly Room at Columbia University, chiefly, Alfred H. Sturtevant (1891–1970), Hermann J. Muller (1890–1967), and Calvin B. Bridges (1889–1938). They co-authored the book *The Mechanism of Mendelian Heredity* in 1915. Morgan received The Nobel Prize in Physiology and Medicine in 1933 for the work: "By conducting statistical studies of the way genetic traits are passed on in fruit flies, Thomas Hunt Morgan broke new ground in genetics during the first decade of the 20th century. His work confirmed that genes are stored in chromosomes inside cell nuclei. He came to understand that genes are organized in a long row inside chromosomes and how traits related to each other correspond to genes that lie close to one another on the chromosomes. He also discovered the crossover phenomenon, in which parts of different chromosomes can trade places with one another." Technically, the last sentence should have been credited to Janssens, but Morgan's group did provide the biological experimental support for the chiasmatype theory.

1.3.4 "EVERYTHING ELSE MAY BE DERIVED FROM THIS LAW"

Another historical reference of important relevance to revisit was the first use of "Mendel's Law(s) of Genetics." It is clear that in the 1865 publication, Mendel did not define such laws, although he was seeking the laws that govern how the differentiating characters appear in the generations from the first hybrid. The Dutch botanist Hugo Marie de Vries published two papers in 1900, one titled *On The Law of Disjunction of Hybrids* and the other *The Law of Segregation of Hybrids.* In the same year, the German botanist Carl Correns published *G. Mendel's Law Concerning The Behavior of Progeny of Varietal Hybrids.* In this paper, Correns stated: "In the hybrid reproductive cells are produced in which the anlagen for the individual parental characteristics are contained in all possible combinations, but both anlagen for the same pair of characters are never combined. Each combination occurs with approximately the same frequency"; and "This I call Mendel's law. It includes the "loi de disjonction" of de Vries, also. <u>Everything else may be derived from this law</u>." Once again, I added the underscore. The itemization of the law of genetics into "multiple laws" first appeared in Morgan's 1919 book *The Physical Basis of Heredity.* These included segregation and independent assortment as the two fundamental principles of heredity and four other principles known as linkage, the linear order of the genes, interference, and the limitation of the linkage groups. Morgan particularly justified the use of the word law in the "six laws of heredity."

A quoted remark from A. J. Bennett in a 1964 paper titled *Mendel's Laws?* pointed it out the most bluntly: "Mendel's Laws, then, is a misnomer ... The laws were read back into his work and have continued to be read back in textbooks ever since." One heavy emphasis in modern genetics books and courses when describing or lecturing on the "Mendel's Laws of Genetics" was the 9:3:3:1 ratio in the second generation (F2) when hybrids of two trait sets are self-crossed, shown in a 16-space Punnett Square. This is used to illustrate the "second law," "independent assortment," and often if not all the time, attributed to Mendel. But this was clearly "read back into his work."

The use of the Punnett Square to show the "independent assortment" of the "differentiating characters" of two pairs works well after the validation of the Sutton-Boveri chromosome hypothesis and the refinement of cytogenetic techniques to clearly observe the chromosome segregation and reduction in the formation of the reproductive cells during meiosis. The meiotic divisions reduce the chromosome numbers by half, and the union of the male and female reproductive cells regenerates the proper chromosome numbers. The later findings and information allowed the designation of the differentiating factors by a pair of letters in the constant parents, caps for the dominant and smalls for the recessive to indicate diploidy, and a single letter in the reproductive cells to indicate haploidy. Mendel reached his conclusions from the hybrid to hybrid crosses without the benefit of knowing diploidy and haploidy. In Mendel's experiments 1 and 3, the self-fertilized second generation needed to produce a ratio of 1:1:1:1 to unequivocally establish the random assortment (**Figure 1-2**), that is, independent assortment.

It makes little difference if we use the 16-space Punnett Square with the 9:3:3:1 ratio or Mendel's simpler 1:1:1:1 ratio to demonstrate the independent assortment of two pairs of factors. The framing of the "second law of genetics of independent assortment" by the 16-space Punnett Square takes attention away, though, from experiments 2 and 4 in Mendel's work. These were back-crosses between the hybrid and the double-recessive constant parent. Experiments 2 and 4 in Mendel's work are actually the true essence of genetic studies of today. This scheme is the one used for "linkage analysis" and will be addressed in the next chapter on human genetics.

Genetics was coined in 1905 by William Bateson, in that letter to Adam Sedgwick, with the comment "Either expression clearly includes variations and the cognate phenomena." The variations and cognate phenomena would include the genetic terms "gene" and "genotype," and further "phenotype." "Gene" came later than "genetics," and its origin was the German word Gen, used in 1909 by the Danish botanist Wilhelm Ludvig Johannsen (1857–1927) to describe the unit of heredity. Johannsen was also the first to use phenotype and genotype to describe the observed and intrinsic differences among the individual representatives of an organism. The term "genome" came from the German word Genom, combined from the German words Gen and Chromosom in 1920 by the German botanist Hans Winkler (1877–1945) to refer to the entire set of chromosomes of an organism. Today, genome refers strictly to the total DNA in all the chromosomes of an organism, often, in the haploid sense, as in "a complete copy."

The origin of genetic studies was Mendel's curiosity into the disappearance of a recessive character in the hybrid and the reappearance of it in the next generation. He carefully examined the contributions of the egg and pollen, that is, the reproductive cells. The reproductive cells are called gametes, first used by the Polish-German botanist Eduard Adolf Strasburger (1844–1912) who studied plant cell mitosis, which he described as "new cell nuclei can only arise from the division of other nuclei." The root words were Greek, "gametḗ" and "gamétēs," literally "husband" and "wife." It must be emphasized that the "variations and cognate phenomena" for genetics also include the passage of the genetic material from a dividing somatic parent cell to the newly divided daughter cells.

BIBLIOGRAPHY

1. https://biobeat.nigms.nih.gov/2024/04/what-is-genetics/
2. https://www.genome.gov/genetics-glossary/Genetics
3. https://www.cdc.gov/scienceambassador/nerdacademy/glossary.html
4. https://onlinelibrary.wiley.com/doi/book/10.1002/9781118269275
5. https://dnalc.cshl.edu/view/16195-Gallery-5-William-Bateson-Letter-page-1.html
6. https://dictionary.cambridge.org/us/dictionary/english/type
7. https://www.genome.gov/genetics-glossary/genotype
8. https://www.genome.gov/genetics-glossary/Phenotype
9. https://www.genome.gov/genetics-glossary/Genome
10. https://www.genome.gov/genetics-glossary/Locus
11. https://www.gov.cn/xinwen/2020-06/01/content_5516649.htm
12. https://www.genome.gov/25520244/online-education-kit-1909-the-word-gene-coined
13. Bateson P. William Bateson: a biologist ahead of his time. *Journal of Genetics.* 2002 Aug;81(2):49–58.
14. Bennett AJ. Mendel's Laws? *School Science Review.* 1964;46:35–42.
15. Correns CG. Mendel's law concerning the behavior of progeny of varietal hybrids. *Genetics.* 1950 Sep;35(5 2):33–41.
16. Fokin SI. Otto Bütschli (1848–1920). Where We Will Genuflect? *Protistology.* 2013;8(1):22–35.
17. Koszul R, Meselson M, Van Doninck K, Vandenhaute J, Zickler D. The centenary of Janssens's chiasmatype theory. *Genetics.* 2012 Jun;191(2):309–17.
18. Maderspacher F. Theodor Boveri and the natural experiment. *Current Biology.* 2008 Apr 8;18(7):R279–86.
19. Manoli I, Fryssira H. Medical genetics and genomic medicine in Greece: achievements and challenges. *Molecular Genetics & Genomic Medicine.* 2015 Sep;3(5):383–90.
20. Martins LA-CP. Did Sutton and Boveri propose the so-called Sutton-Boveri Chromosome Hypothesis? *Genetics and Molecular Biology.* 1999;22(2):261–272.
21. Mendel G. 1866. Versuche über Plflanzenhybriden. Verhandlungen des naturforschenden Vereines in Brünn, Bd. *IV für das Jahr* 1865, Abhandlungen, 3–47.
22. Monaghan F, Corcos A. On the origins of the Mendelian laws. *Journal of Heredity.* 1984 Jan–Feb;75(1):67–9.
23. Morgan TH. Chromosomes and heredity. *The American Naturalist*, 1910;44(524):449–96. http://www.jstor.org/stable/2455783
24. Sutton WS. On the morphology of the chromosome group in Brachystola magna. *Resonance* 2011;14:398–411.

2 How Special Is Human Genetics?

How many children do you wish to have?

Genetics as a special branch of biology was born out of the studies of plants such as the garden peas and insects such as the fruit flies. The law of genetics, which we redefined as the distribution of the parental chromosomes in the haploid gametes by *segregation*, that is, one of either of each pair of chromosomes from the diploid pregametic parent cell. Plant genetics and insect genetics are still in practice today on more advanced levels following the "molecular revolution" in biology, with the advent of gene cloning and genome sequencing. Florida Cancer Specialists & Research Institute runs an XM radio ad that boasts their use of "next-gen sequencing" to bring "personalized medicine" to the treatment of cancers. This would be the cutting edge of human genetics of today in the post-genomic era after the completion of The Human Genome Project, officially announced on June 26, 2000.

It was always advertised that The Human Genome Project would give us the complete code book of the human form of life. With the complete human genome sequence information, we would have recognized every human gene and understood the biological functions of its gene product. If we use a metaphor of automotive repairs, it would be as if for the recorded human history of medicine from somewhere in 3000 BCE to before June 26, 2000, we had been performing the repairs without the instruction manual. Now the completed and gap-filled human genome sequence of the 2022 edition is that instruction manual. But do we know how to read it?

A catalog of human hereditary disorders is updated at the Online Mendelian Inheritance in Man (OMIM), a database created in the early 1960s by Dr. Victor A. McKusick (1921–2008) as Mendelian Inheritance in Man (MIM) and moved online in 1985 jointly by the National Library of Medicine and the William H. Welch Medical Library at Johns Hopkins University. As of the May 21, 2024, update, the OMIM entry statistics listed the following:

TABLE 2-1. OMIM entry statistics (updated on May 21, 2024)

	Autosomal	X Linked	Y Linked	Mitochondrial	Total
Gene description	16,418	770	51	37	17,276
Gene and phenotype, combined	21	0	0	0	21
Phenotype description, molecular basis known	6,425	386	5	34	6,850
Phenotype description or locus, molecular basis unknown	1,388	110	4	0	1,502
Other, mainly phenotypes with suspected Mendelian basis	1,633	100	3	1	1,737
Total	25,885	1,366	63	72	27,386

Table 2-1 is very rich in information and also very rich in the lack of information. We begin dissecting the richness of information and the lack of information and set the goal for this book to serve as a reader's manual to the book of human life, the instruction manual of how to maintain and repair our form of life.

2.1 The Genetic Basis of Physiological Differences—The Example of Alcohol Metabolism

2.1.1 ALCOHOL METABOLISM ENZYMES AND GENES

For the entries listed as "gene description," we can look at the genes that code for the enzymes involved in metabolizing alcohol in the body. The first enzyme is alcohol dehydrogenase (ADH).

There are seven genes in the ADH family, located on human chromosome 4 at 4q22-23, coding for five classes of ADH. The class I enzymes, formed by two of three different subunits, alpha, beta, and gamma, have a high affinity for ethanol and are responsible for over 70% of the ethanol metabolism in the liver. They are encoded by three different genes, *103700, alcohol dehydrogenase 1A (*ADH1A*), *103720, alcohol dehydrogenase 1B (*ADH1B*), and *103730, alcohol dehydrogenase 1C (*ADH1C*), in a tandemly arranged cluster. ADH1A is made in the fetal and infant liver, while ADH1B and ADH1C are made in the adult liver. ADH catalyzes an oxidative reaction, removing two hydrogen atoms to convert alcohol to acetaldehyde, that is, from the hydroxyl group to a carbonyl group. The main organ that metabolizes the ingested ethanol is the liver, but a small amount can be metabolized in the brain, pancreas, and gastrointestinal tract.

An aldehyde dehydrogenase in a second oxidative reaction catalyzes the removal of another hydrogen atom and the addition of a hydroxyl group (from a water molecule) to convert the acetaldehyde to acetic acid (acetate). The final product acetate is converted by acetyl-CoA synthetase to acetyl-CoA, which can directly enter the citric acid (TCA) cycle, a cycle of biochemical reactions that metabolize food and produce energy. In the liver, there are two highly homologous ALDH enzymes responsible for converting acetaldehyde: ALDH1, in the cytosol, and ALDH2, in the mitochondria. They are encoded by the *ALDH1A1* gene, *100640, and the *ALDH2* gene, respectively. The *ALDH2* gene, +100650, is an entry in "gene and phenotype, combined."

2.1.2 POPULATION DIFFERENCES IN ALCOHOL METABOLISM—PHENOTYPES

Beverage alcohol, specifically ethanol, has multiple toxicological effects on the body, causing interference of brain functions, heart and liver damages, and increased risks of cancer with either binge or long-term drinking. Acetaldehyde, the product of the first step of ethanol metabolism, is also toxic, and its accumulation in the body can cause highly aversive reactions such as facial flushing, nausea, and increased heart rate.

The first step in ethanol metabolism, the conversion to acetaldehyde, is the slower one, known as the "rate-limiting" step in biochemistry. This would make sense as the reaction's toxic byproduct acetaldehyde can be removed by the second step, the conversion to acetate, which is essentially good food. Overdrinking would obviously overwhelm the two-step metabolic pathway and lead to the accumulation of both ethanol and acetaldehyde in the body. For some ethnic groups, such as the East Asians, including the Chinese, the Japanese, and the Koreans, drinking a small amount of alcohol can lead to a quick accumulation of aldehyde and the aversive reactions for the majority of the population.

2.1.3 THE GENETIC BASIS OF POPULATION DIFFERENCES IN ALCOHOL METABOLISM—GENOTYPES

The reason behind such a "human difference," or, more precisely, "phenotype," that is, manifested difference in alcohol metabolism, is the multiple "genetic isoforms" of the alcohol metabolic enzymes. The ADH1A, ADH1B, and ADH1C isoforms are "biochemical isoforms" encoded by independent genes, which are believed to have evolved from gene duplication events. They are greater than 90% identical in their amino acid sequences. The genetic isoforms, however, are encoded by different alleles, from the Greek word "allelemorph = of another form," at the same gene locus. The genetic isoforms are more often referred to as "variants," which can refer to the gene alleles or the enzymes. In the case of the ADH and ALDH enzymes, such genetic isoforms can differ by only one single unit of their coding sequences at the DNA level, and changing only one single amino acid residue at the protein level. These are called "single-nucleotide polymorphisms" (SNPs). We will have a more extensive look at the utility of these SNPs in a later chapter.

Some of these genetic isoforms or variants have altered enzymatic activities in the reactions they catalyze. For ADH, the *ADH1B*2* variant has the amino acid histidine (His) at position 48 instead of arginine (Arg). The *ADH1B*3* variant has an Arg to cysteine (Cys) substitution at position 369. These single genetic and biochemical changes increase the alcohol to aldehyde turnover rate by 70- to 80-fold. *ADH1B*2* is the most common in East Asian populations, at >40%, whereas *ADH1B*3* is primarily found in populations of African descent, at up to 40%. These variants were associated with a protective effect against alcohol dependence. For aldehyde dehydrogenase, the best studied is the *ALDH2*2* variant, having a glutamic acid (Glu) to lysine (Lys) substitution at position 504, resulting in an inactive enzyme incapable of oxidizing acetaldehyde to acetate and in turn the accumulation of the toxic acetaldehyde. *ALDH2*2* is very common in the East Asian populations (Chinese, Japanese, and Korean), at 45%, but nearly absent in European or African populations. This is the main reason that the East Asian populations display the aversive effects such as severe facial flushing, nausea, and increased heart rate even when consuming a small amount of alcohol. Having just one *ALDH2*2* allele (while the other one is "normal") has a strong protective effect against alcohol dependence. We can view the protective effect of the faster ADHs (*ADH1B*2* and *ADH1B*3*) as a result of faster removal of ethanol, provided that the aldehyde dehydrogenase at the next step functions normally. We can then view the protective effect of the inactive aldehyde dehydrogenase (*ALDH2*2*) as a result of triggering a faster aversive response, turning people away from drinking.

2.2 The "Forward Genetics"—From Protein Information to Gene Information

2.2.1 THE FORWARD JOURNEY BEGINS WITH THE PURIFICATION OF A PROTEIN (ENZYME)

The "genetic" information that we have on the population differences in the alcohol metabolism enzymes was the outcome of a paradigm known as "forward genetics." It involves the clinical identification of a phenotype, often referred to as a "disorder," such as an inability to effectively metabolize alcohol, for which the genetic basis is unknown at the time. The next steps involve biochemical analysis to reduce the physiological phenotype to a metabolic pathway, and further to the specific enzymes. When the source material is available in good quantity and at relatively low costs, the enzymes are purified to homogeneity and their amino acid sequences are determined. In addition, the purified enzymes are used to immunize an animal host, such as a rabbit, to raise specific antibodies against the enzymes.

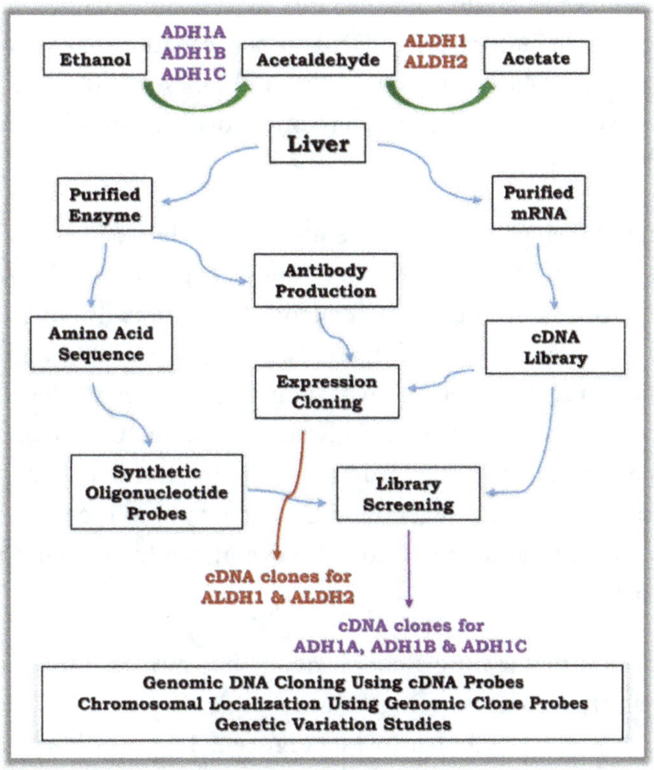

FIGURE 2-1. Forward genetics in the cloning of the genes encoding enzymes in alcohol metabolism

The amino acid sequence can be "back-translated" to the messenger RNA (mRNA) sequence, but with "redundancy" or "degeneracy." For example, a triplet codon of GGG, GGA, GGU, or GGC would specify a glycine (Gly). But a Gly could be encoded by any of the four different codons. Francis Crick (1916–2004) defined the central dogma of molecular biology as the "residue-to-residue" genetic information transfer, which is unambiguous and nonredundant from nucleic acids to nucleic acids, and from nucleic acids to proteins. The second step of information transfer is technically still to nucleic acids on the anticodons in the transfer RNA (tRNA). But he emphasized that the information transfer cannot be from proteins to nucleic acids in nature. Thus, "back-translation" does not occur in nature and is a human exercise on paper to predict the gene sequence from a protein sequence.

In the amino acid sequence, a region of the least redundancy would be selected to guess the corresponding mRNA sequence. A synthetic pool of short DNA, known as oligonucleotides, representing all possible sequences are then made for use as probes to screen a "complementary DNA (cDNA) library," the collection of DNA copies of (nearly) all mRNA from a tissue or cell source. In the examples of the alcohol metabolic enzymes, the material source for protein purification is the liver, or hepatocytes grown in culture. The cDNA library can also make the "recombinant human proteins" in the *Escherichia coli* host cells so that the antibodies raised against the purified enzymes can be used to screen for which bacterial colony (clone) hosts the specific copy of the gene (mRNA) encoding the purified enzyme.

A successful screening, either with the degenerate oligonucleotide probes or with the antibody probes, would give us a "cDNA clone," a double-stranded copy of the mRNA encoding the protein of our interest. The cDNA is then completely sequenced, traditionally with "Sanger's dideoxy chain termination" method. A large amount of the cDNA is then prepared and labeled as a probe to screen a "genomic DNA library," the collection of (hopefully but never) all genomic DNA fragments of the organism. A successful screening would give us a "genomic DNA clone." The cloned genomic DNA would then be completely sequenced and compared to the cDNA sequence to determine the exon-intron junctions. The cloned genomic DNA containing the gene of interest would then be used as a probe to localize the gene by "fluorescence in situ hybridization" (FISH) to the metaphase-condensed chromosomes.

The "forward genetics" journey for the alcohol metabolic enzymes has been completed, or their genes "cloned." We can think of it as going forward to getting the final resolution of the genetic information on a clinically described phenotype. **Figure 2-1** gives a flowchart view of the forward genetics paradigm using the examples in the cloning of the genes encoding the enzymes in alcohol metabolism. A good number of genes were cloned by forward genetics, the first being that of β-globin. Other commonly well-known genes cloned by forward genetics were insulin and angiotensin

I–converting enzyme (ACE) (OMIM entry +106180), a key regulator of blood pressure and electrolyte balance. ACE catalyzes the second step in the generation of the potent vasoconstrictor angiotensin II, from angiotensin I, released from angiotensinogen by the aspartyl protease renin. The renin-angiotensin system (RAS) is a major network of enzymes and effectors that regulate blood pressure and the target of antihypertensive drugs such as the ACE inhibitor captopril.

2.2.2 THE MOST DIFFICULT STEP IS THE FIRST STEP

The steps required to complete a forward genetics cloning of a gene, especially a human gene, are not always possible, starting from almost the first step. The examples we looked at above were all "abundant proteins" from a tissue/cell source that can be secured at a reasonable cost. One of the most popular sources was the pancreas, porcine or bovine, cheap and rich in many enzymes and hormones. Commercial medicinal insulin was first extracted from the porcine or bovine pancreas. It was actually the high cost and effort involved in the extraction of insulin from such sources that incentivized the molecular cloning of its genes (cDNAs), first of other mammalian species and then of humans. Insulin has another favorable characteristic as being a small protein of two relatively short polypeptide chains. The amino acid sequences of insulins from various species were readily determined using classical biochemical methods. The animal-sourced insulin had issues of impurity that caused clinically significant problems. In the years before the molecular biology revolution, efforts were made to chemically synthesize this hormone. The first "recombinant" human insulin produced by *E. coli* cells actually used a chemically synthesized "human insulin gene" cloned in a proper vector, again, because the coding sequences of the short insulin peptides are also short.

The eventual cloning of the human insulin cDNA relied on the mRNA from human insulinoma cells grown in culture, using a rat insulin cDNA as the probe. This practice was very common as the genes of interest were highly conserved across the species. The cloning of the β-globin gene benefited from a naturally occurring "signal-to-noise ratio" that was not replicated anywhere else. The rabbit reticulocytes, precursors of the red blood cells, make so much α- and β-globin for the eventual hemoglobin in the red blood cells that 90% of the total proteins were of these two kinds. The heavy polysomes contained almost all β-globin mRNA at >90%, while the mRNA in the light polysomes was >70% of α-globin. Converting these mRNA into the cDNA and cloning them into a proper vector (plasmid) for propagation and characterization were thus not challenging at all. The β-globin mRNA abundance at >90% in the heavy polysomes means that once they are converted to the cDNAs in the plasmid vector and plated onto an agar petri dish, randomly picking 10 clones would give us nine with the β-globin sequence. We can still view the random picking as screening, as in, finding out how many in a sample number of clones are of the kind we want.

2.2.3 THE COMPLEMENTARY DNA (MESSENGER RNA) BYPASS ENHANCES THE SIGNAL-TO-NOISE RATIO

A principle that was demonstrated by the globin mRNA/cDNA cloning was the amplification of the gene signal at the transcriptional level. Most current and modern textbooks neglect to emphasize the quantitative aspects between the amount of a specific mRNA *per* cell and the copy number of a gene, at a fixed 2 *per* cell. Depending on the specific gene and protein, its mRNA can range from several to several thousand copies *per* cell. It is also important to note that not every gene is expressed in a specific cell type. Thus, the representation of the gene at the mRNA/cDNA level is much more advantageous than the two copies *per* cell, or one copy *per* haploid genome.

It is thus not surprising that in the forward genetics route, there was not a documented success in directly cloning a genomic DNA fragment of a gene with the degenerate oligonucleotide probes in a genomic library screening. The degree of degeneracy is a big compromise of the necessary probe signal and to screen a genomic DNA library, the entirety of the haploid genome contributes to the noise.

In the examples listed above, the cDNAs of the human ADH isozymes were cloned by using either the antibodies or the degenerate oligonucleotides as the probe to screen a very popular human liver cDNA library of the times, prepared by Savio L. C. Woo of The Howard Hughes Medical Institute (Houston, TX). The cDNAs of the human aldehyde dehydrogenases 1 and 2 were cloned in a similar manner. These enzymes are at sufficient abundance levels and can be tracked by a functional assay to allow their purification from a natural source, such as a liver autopsy sample. As for human ACE, the enzyme was purified from the cortex of a fresh postmortem human kidney, a partial amino acid sequence was determined, and used to design a degenerate oligonucleotide probe for screening a human umbilical vein endothelial cell cDNA library.

2.2.4 GENETIC INFORMATION ENABLES SCREENING OF DIFFERENCES IN POPULATIONS

With these classical examples, the picture is very clear for the practice of forward genetics to determine the gene encoding an enzyme, for which a genetic variant may cause a physiologically manifested phenotype. Such was the case of β-thalassemia, which is a disease associated with a reduced production of β-globin, and of the sickle cell disease (SCD)/anemia, which has a single-nucleotide/amino acid substitution in the β-globin sequence at either the DNA or the protein level. With the knowledge and information on the genes encoding these phenotype-associated variants, it is possible to ask the question of how many persons in a population may be affected or may be a carrier of such a genetic variant. The SNP in the SCD was the basis of the very first molecular diagnostic test. At first, it involved using human genomic DNA in a classic Southern blot analysis, but the invention of the polymerase chain reaction (PCR) greatly improved its sensitivity and user-friendliness in practice in the mid-1980s.

On the other hand, it would be very impractical and costly if we rely on biochemical analysis to screen a population. For the β-globin diseases, the source material is the red blood cells, which can be readily obtained. But the biochemical analysis would still be costly and time-consuming, as well. For the alcohol metabolic enzymes, the source material would be a not so small slice of the liver.

2.3 The "Reverse Genetics"—Getting the Genetic Information First

The term *reverse genetics* was initially used to describe the studies of phenotypic changes associated with artificially changing the gene sequence, via genetic engineering techniques such as "site-directed mutagenesis," once the gene information was obtained via forward genetics. It also included the studies of phenotype changes in animals expressing a (human) transgene. Such efforts were in a broader sense made to fulfill the Koch Postulates, as they apply from microbiology to molecular biology or molecular genetics. The idea was that, if the starting point of forward genetics was the clinical description of a phenotype (disease), and the gene information was obtained via reducing the complexity of the phenotype (disease) to a single pathway, biochemical reaction, and enzyme, by manipulating this single point at the gene or protein level the phenotype should be recapitulated. And if not, something must have gone wrong along the way.

The genetic manipulation can be the modification of a gene, either from a normal allele to a mutant to reproduce a genetic disorder or from a mutant to the normal to cure a disease. The Nobel Laureate Oliver Smithies (1925–2017) shared the 2007 prize in physiology and medicine for his work of using homologous recombination to modify genes in mice, motivated by his desire to cure sickle cell anemia. Today, a more precise technology known as the CRISPR/Cas9 can be applied to genome editing, including in humans. In December 2023, the U.S. Food and Drug Administration (FDA) approved the first gene therapy using this technology for patients with sickle cell. On the contrary, gene targeting or knockout of the mouse angiotensinogen gene resulted in an expected phenotype of hypotension, as the source of the vasoconstrictor angiotensin II is now removed.

There were and still remain a good number of hereditary disorders in humans for which the genetic basis could not be determined via forward genetics, starting with the purification of the relevant protein molecule. For some of these, the candidate protein molecule was difficult or impossible to identify, and for others, the candidate protein exists transiently when performing its function, not accumulating in a cell or tissue source at enough quantities to allow any attempt at purification.

The best known examples of these are alkaptonuria (AKU), cystic fibrosis (CF), and Huntington disease (HD). AKU was the first ever hereditary disorder to be characterized in 1902, as a recessive disease, by Sir Archibald Garrod (1857–1936). The recessive inheritance of CF was first described in 1949 by Dorothy Andersen (1901–1963). George Huntington (1850–1916) documented the pattern of familial inheritance of HD, in 1872, a few years after Mendel's landmark paper and decades before the "rediscovery" of Mendel's work and the framing of the law of genetics. HD is an autosomal dominant disorder.

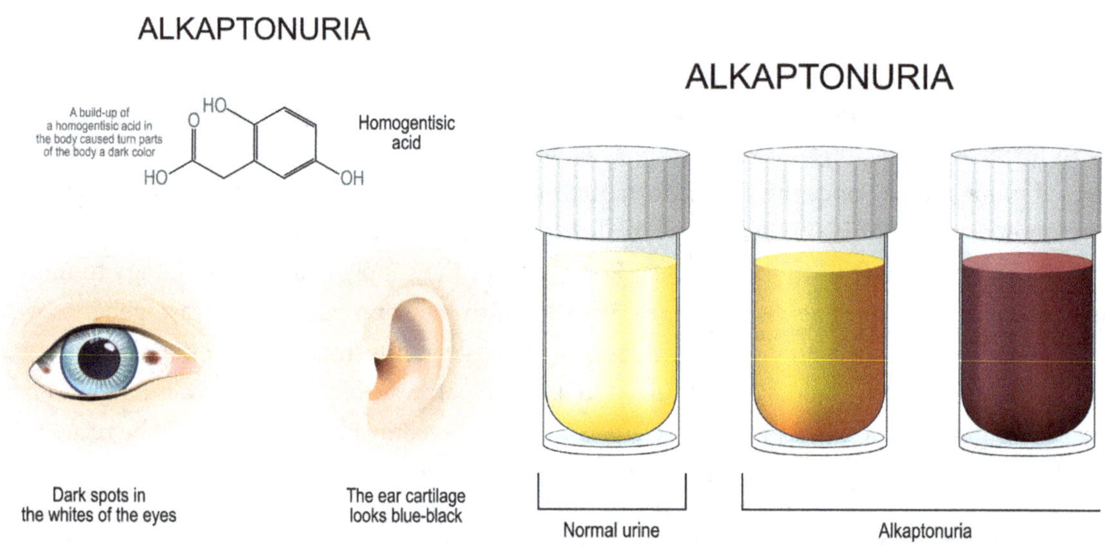

FIGURE 2-2. Alkaptonuria

2.3.1 ALKAPTONURIA—FROM 1902 TO 1993

The example of AKU probably captured the best of all the struggles and ingenuities in the field of human genetics. Evidence of this disease was found in the Egyptian mummy Harwa of 1,500 BC. The early clinical description of the disease was mainly the presence of a high amount of homogentisic acid (HGA) in the patient's urine, a metabolite of tyrosine and phenylalanine. HGA turns black upon exposure to air and is used for the diagnosis of patients from childhood (**Figure 2-2**). Between 1902 and 1908, Sir Archibald Garrod, with advice from William Bateson (he was everywhere!), proposed a single-gene recessive defect based on the analysis of inheritance patterns in consanguineous families, with a biochemical defect of an enzyme that breaks down HGA, the homogentisic acid oxidase (HGD). Fifty years later, a liver biopsy sample of an AKU patient allowed the determination of HGD being the only defect in the tyrosine metabolism with comparisons to the non-AKU liver preparations. Taking advantage of the allele sharing in consanguineous families, a method called "homozygosity mapping" was applied to localize the "disease gene" to human chromosome 3 at 3q2, displaying coinheritance with the recessive familial hypocalciuric hypercalcemia (FHH). The FHH gene was mapped to 3q a year prior by "linkage analysis" using "polymorphic markers."

There is a lot to unpack in the short summary of the literature regarding the "mapping" (localization) of the candidate HGD gene to human chromosome 3 at 3q2. Fifty years after the recognition of its Mendelian recessive mode of inheritance in humans and the proposed enzyme that is missing in the AKU patients, the first meaningful biochemical attempt at characterizing the enzyme remained a negative success. Yes, something and the only thing is missing in the pathway of tyrosine metabolism, but we could not identify what exactly. For insulin, on the other hand, rich in the pancreas, the pure form in large quantities was achieved in 1921. The digestive enzyme trypsin was purified by crystallization from the pancreas in 1930. Clearly, there were challenges that could not have been overcome in the efforts to obtain the pure form of the HGD enzyme. The route of forward genetics to identifying and cloning the HGD gene had a hard stop from the starting line.

2.3.2 ALKAPTONURIA—FROM GENETIC INFORMATION TO PROTEIN INFORMATION

The mapping of the HGD gene relied on the prior known location, or, landmark of the FHH gene. Thus, we have a situation that the genetic information was more or less obtained before the protein information. The protein information on the HGD enzyme came first in the form of a purified mouse counterpart, in 1995, 2 years after the human gene localization. This work performed a comprehensive biochemical characterization of the mouse HGD enzyme, providing information on the molecular weight, isoelectric point, optimal pH for enzymatic activity, and enzyme kinetics and cofactors. No amino acid sequence was reported. An attempt in 1968 to purify the HGD enzyme from the bovine liver resulted in a much more crude preparation to allow only some preliminary enzymatic activity studies.

The breakthrough on the human HGD enzyme came by way of a rather large detour, via the cloning of the HGD (HGO) counterpart, *hmgA*, from the fungus *Aspergillus nidulans*. We will omit the discussion or description of the details involved in the fungal *hmgA* gene cloning, except to say that the confidence was based on the functional parallel in the metabolism of amino acids carrying a benzene ring, such as phenylalanine and tyrosine. Our emphasis will be given to the fact that the cross-species conservation of the enzymes in the pathway permitted the identification of the human HGD cDNA from the expressed sequence tags (ESTs). ESTs represented a revolution in the strategic thinking about decoding the human genome. Instead of a targeted screening, the entire mRNA (cDNA) pool of a tissue or cell type was randomly sequenced and then cataloged for many tissues or cells. This was enabled when the cost of DNA sequencing was significantly reduced and the technology significantly improved in the early 1990s.

The human HGO cDNAs allowed the complete deduction of its amino acid sequence, the determination of the tissue-specific expression, the isolation of genomic DNA clones containing its gene, and the localization of the gene to a more precise location on human chromosome 3 at 3q21-23. More importantly, it provided the sequence information to develop molecular diagnostic methods for genotyping AKU patients in families to confirm the mutations. Further, the variant in the patients, Pro230Ser, was confirmed as a loss-of-function allele.

On the topic of "human genetics," in the long journey from 1902 to 1995 of resolving the molecular basis of AKU in humans, there were two important events we must highlight. The first is the coinheritance of AKU and FHH, and the second is the localization of the FHH gene to human chromosome 3q via "linkage analysis."

2.4 Linkage Analysis—Translating the Fruit Fly Genetics to Human Genetics

The term *linkage* has two meanings in genetics: first, genes on the same chromosome are literally linked by the linear DNA material in the chromosome. Thomas Morgan initially rejected the Sutton-Boveri chromosome hypothesis of heredity due to the fact that the identifiable traits in the model organism in his lab, the fruit fly *Drosophila melanogaster*, greatly outnumber the pairs of chromosomes. Morgan then realized that some of the traits can be carried by the same chromosome pair and called this "Mendelize together." It is not clear or important to find out how prevalent "to Mendelize together" had been used, but today, the proper genetic term is linkage.

2.4.1 TRAITS ON DIFFERENT CHROMOSOMES MENDELIZE (ASSORT) INDEPENDENTLY

Opposite of "to Mendelize together" would be exactly "to Mendelize independently," which is also exactly the law of genetics. In **Chapter 1,** we reviewed the work performed by Mendel and one critical piece was the cross between the double heterozygote hybrid of round/wrinkled-yellow/green with the double recessive homozygote parent, wrinkled/green (**Figure 1-2**). This produced four different progeny phenotype combinations, two of the parental types, round/yellow and wrinkled/green, but also two that were different from the parent types, round/green and wrinkled/yellow, at a ratio of 1:1:1:1. The new combinations are now referred to as "recombinants," literally, the recombination of the parental traits, performed during gametogenesis by the hybrid with the four elements. We know today that the yellow/green and round/winkled are carried by two different pairs of chromosomes, 1 and 7. By the Sutton-Boveri chromosome hypothesis, the four possible combinations would be presented at the 1:1:1:1 ratio, as Mendel noted himself in his 1865 paper. To express this in another way, since half of the combinations are new, and are the "recombinants," we say that the recombination frequency is 50% or the recombination fraction is 0.5.

It is important to emphasize that the business of "genetics" is done and completed at the end of gametogenesis (meiosis), but no method (not even very much today without sacrificing the gametes) is available to know what happened in terms of how the chromosomes, or trait elements were distributed in the eventual gametes. Progenies are required to "report" on this distribution, and the best experimental setup would be to cross with a double recessive phenotype (homozygote) so that the hybrid gamete trait elements are fully presented phenotypically without being masked in the progeny. This was done by Mendel with experiments 2 and 4 highlighted in **Figure 1-2**, decades ahead of the Sutton-Boveri chromosome hypothesis. It is the simplest cross, and at the same time, it provides both the necessary and sufficient information to make a genetic determination.

2.4.2 TRAITS ON THE SAME CHROMOSOMES CAN MENDELIZE (ASSORT) TOGETHER

In the fruit fly, if the traits carried on the same pair of chromosomes truly Mendelize together, due to the physical material linkage, one array of each trait pair would stay together and the other array would, too, at the same time. If a hybrid is crossed with a double recessive, the progeny would present only two combinations, reproducing the parental phenotypes and there would not be any recombinants.

What did Morgan's lab observe in their genetic experiments using the trait pairs of fruit flies? Some traits behaved just like the traits Mendel studied in the garden peas, on a hybrid to double-recessive cross, like those shown in experiments 2 and 4 in **Figure 1-2**, the four combinations appeared at a ratio of 1:1:1:1. In other words, there was no "co-Mendelization." They produced the recombinants at RF = 50% (0.5). For some traits, the ratio departed from 1:1:1:1, with the recombinants at RF < 50% (0.5). For these traits, there is "co-Mendelization," but not completely.

In a pregametic cell, each chromosome is duplicated (with its DNA replicated, two different words, same meaning) and each pair with the four chromatids would be distributed in the four eventual gametes following the meiotic divisions. In Mendel's experiments, he looked at traits on seven different pairs of chromosomes, so the outcome would always be 1:1:1:1, with 50% being the recombinants in a cross involving a double heterozygote and a double recessive. In the flies, if some traits are carried by the same pair of chromosomes, the "physical linkage" would prevent the recombinants. For example, for a string of six trait sets, on one of the pair, it is "ABCDEF," and on the other, it would be "abcdef," as shown in **Figure 2-3** (shown as **Figure 1-3** in **Chapter 1**). The two homologous chromosomes are duplicated to give two copies of "ABCDEF" and two copies of "abcdef," but only two types of gametes, "ABCDEF" and "abcdef." If that were the case, the expected recombination frequency would be zero.

The observed recombination frequencies for some of the trait sets, however, were between zero and 50%. Can chromosomal theory and co-Mendalization be reconciled with these experimental observations? Crossing-over between the non-sister chromatids before the first division in meiosis provides the answer (**Figure 2-3**). In a typical figure illustration, the two duplicated homologous chromosomes, each with the two sister chromatids, stand side-by-side. The two inner non-sister chromatids would cross over and, via a series of biochemical reactions, undergo an exchange of the chromosome material. The two outer non-sister chromatids are not involved.

As a result of the crossover and the exchange of the chromosome material between the inner non-sister chromatids, the four gametes may end up with four different chromosomes. From the outer two non-sister chromatids are the nonrecombinant "ABCDEF" and "abcdef," and from the inner two non-sister chromatids are the recombinant "ABCdef" and "abcDEF." The latter two are recombinants over the length of the chromosome. But if the points of interest are in the "ABC"/"abc" range, the observed phenotypes would only be two, "ABC" and "abc," and there is no recombinant type. The same is true for "DEF" and "def." It is not difficult to see that between "Aa" and "Ff," there are the maximum chances of crossover, hence the maximum chances of recombinants. Between any two adjacent letters, the chances are much smaller. Depending on the actual positions of "AaBbCcDdEeFf" on a liner chromosome, the crossover and exchange between two loci can be so frequent that the observed recombinants are at RF = 50% (0.5).

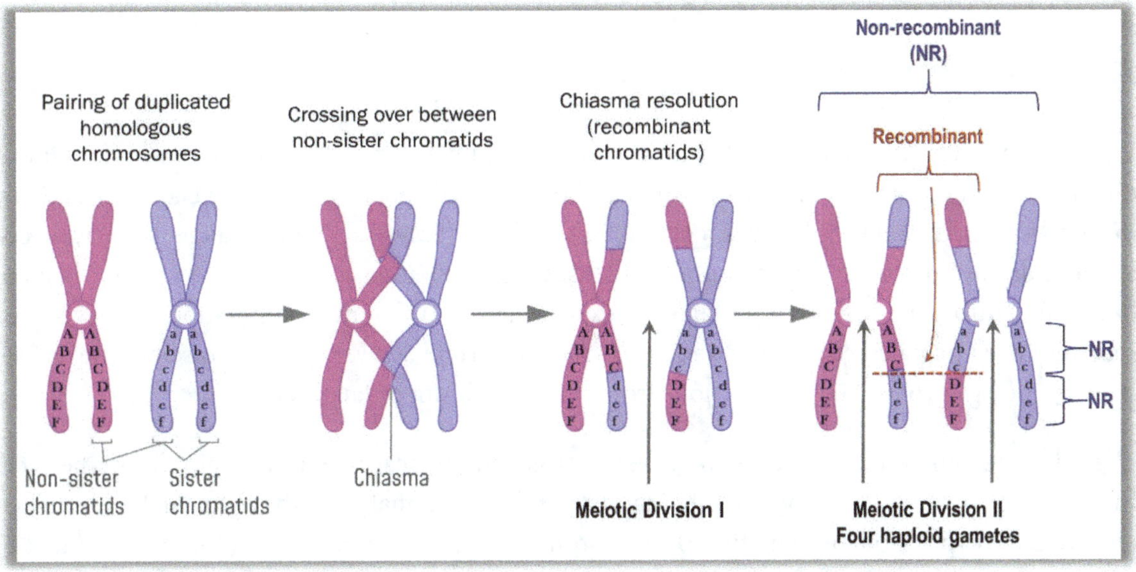

FIGURE 2-3. Meiotic division and allelic recombination across loci on the same chromosome

2.4.3 WE RELY ON THE ALLELE RECOMBINATION FREQUENCY/FRACTION TO DETERMINE THE LOCUS–LOCUS RELATIONSHIP—ON THE SAME OR NOT ON THE SAME CHROMOSOME

We have now transitioned our discussion of "genetics" into determining the "positional relationships" of traits/loci in relation to chromosomes. Both "Mendelian Genetics" and "Morgan Genetics" rely on phenotypic observations to infer on the genetic compositions. In two conditions mentioned above, true Mendelian, that is, traits are on different chromosomes, or, traits on the same chromosome but far enough apart, we would observe an RF = 50% (0.5). From the observed RF, we would be unable to determine which one of the two conditions produced the results.

When the observed RF is smaller than 50% (0.5), it can only be that the two traits are on the same chromosome. Distilling from all knowledge and information in the history of genetics from 1865, the essence is "to determine the locus-to-locus relationship," and it is done by "determining the allele–allele relationship across the two loci."

2.5 The First Genetic Map, the First Physical Map, and the Art of Genetic Mapping

I have always used this example (joke) in my course of human genetics, and also molecular biology during the part of introducing the concepts and practices of genetics. I ask the class: "What is the most prevalent human hereditary disorder (disease)?" The responses would range from sickle cell, to muscular dystrophy, and onto complex diseases like Alzheimer and cancer. I go on a step and hint: "This disorder (disease) is suffered by exactly half of the human population." I have never received the anticipated correct answer, or simply, the correct answer. When I disclose the answer, it always brings two immediate reactions from the class, doubt or laughter.

There is no example that captures the essence of genetic (disease gene) mapping more perfectly than this. There are two phenotypes, being male and being female. At the cytogenetic level, there are two karyotypes—the presentation of the chromosomes in a somatic cell, (46,XX) and (46,XY). Whenever XX is present, the person is identified as a female, while XY is identified as a male. Or we can say that maleness is "associated" with the presence of the Y chromosome. If we follow the hypothesis that there was once only the X chromosome and Y came into existence by evolving from the X, then Y can be considered the mutant or the disease variant. This designation would likely generate the least amount of social disturbance, but we will quickly depart from this topic. Yet we should have no resistance to claiming that the first human "disease gene" mapped to a chromosome was maleness to the Y. Of course today we know that the Y chromosome harbors more than one actual gene, and maleness is not a single-gene disease. The difference here, however, is not conceptual but rather an issue of resolution.

The first genetic map was established in a manner entirely not dissimilar to what we have seen above. When the *Drosophila* mutations accumulated to a good number in the laboratory of Thomas Morgan, the researchers began to see linkage groups of the loci harboring the mutations (alongside their normal alleles). Within the linkage groups, the alleles across the loci recombine at frequencies less than 50% if they are at a close distance, but all would recombine with alleles of loci in other linkage groups at the 50% frequency. The lowest hanging fruit of "associating" a linkage group to a specific physical chromosome was that on the sex chromosome X. In the fruit fly, only females have two X chromosomes that would allow for crossing-over, and thus recombinants in the phenotypic manifestations in the progeny. The males have a single X and no crossing-over or recombinants are possible. Whereas the linkage groups on the paired autosomes could undergo such chromosomal events and manifest the outcomes in either sex, more or less equally across the lengths. It is essentially a process of elimination for the X-linked group of loci. The other linkage groups, the "second" and the "third," at the early stages of fruit fly genetic studies (1910 to the late 1920s), remained genetically mapped groups without a specific chromosome assignment.

In the early 1930s, the cytogenetic analysis of the uniquely characteristic salivary gland chromosomes of *Drosophila* provided the breakthrough to assign the other linkage groups of loci to specific autosomes. The word cytogenetics literally means looking into the cells (cyto) for information on genetics (chromosomes). In the salivary gland of the fruit fly and other flies (Diptera), the chromosomes are enlarged and give more structural details. The ability to describe the "bands" in the chromosomes in such details, known as the "salivary chromosome coordinates," allowed geneticists to document the precise points of physical chromosomal changes, either naturally occurring or artificially induced. These coordinates provided a direct way to associate the physical material map of chromosomes to the manifested phenotypes in genetic tests that created the linkage groups, that is, the genetic maps.

The human and other mammalian chromosomes presented much more of a challenge for cytogenetic studies because of the difficulties in the examination of very crowded nuclei containing all the condensed metaphase chromosomes. It was not until the 1950s with the advancement of techniques to process the cells to obtain better metaphase chromosome spreads that the human diploid chromosome numbers were determined to be 46. After this, the low-hanging fruits of hereditary human disorders with large chromosomal abnormalities were harvested. These included the numerical abnormalities such as trisomy 21 in Down syndrome, (45,X) in Turner syndrome, (47,XXY) in Klinefelter syndrome, and translocation events such as the Philadelphia chromosome, involving chromosome material exchanges between 9 and 22. Duchenne muscular dystrophy (DMD) had long been identified as an X-linked recessive disease expected to be manifested only in boys, who have only one copy of the X chromosome. In young girls who must be carriers of a mutated DMD gene on one X chromosome to present DMD-like clinical manifestations, X to autosomal translocations were identified to consistently involve the presumed good X chromosome with breaks at Xp21. This association suggested the location of the DMD gene in this region. Such translocation events would inactivate the normal DMD (dystrophin) gene and present the DMD phenotypes. There are other examples in which cytogenetic work facilitated the localization of genes to the human chromosomes, but clearly this approach is not to be relied on for mutations caused by much smaller, down to the single-nucleotide changes.

2.6 The Similarities and Differences between *Drosophila* Genetics and Human Genetics

2.6.1 THERE IS ONLY ONE BRAND OF GENETICS CONCERNING DIPLOID ORGANISMS

First, the principle law of genetics is the same for fruit flies and humans, in that the homologous chromosome pairs are distributed in the gametes by either one of each pair. But genetics is never studied at the gametic level, and it requires the progeny phenotypes to report back on how the chromosomes were distributed during meiosis (gametogenesis). In both fruit flies and humans, the basic genetic question being asked is the same: On which chromosome is this gene (locus)? In fruit flies, the initial goal was to prove or disprove the Sutton-Boveri chromosome theory of heredity. The first actual gene (mutation) localized to a chromosome was that of the white eye. In the efforts that ensued, several additional mutations (genes) were localized to the X. But in the autosomes, the linkage groups were established for a rather long time before the chromosome assignments. This was possible because all the loci in the linkage groups were phenotypic, with physical manifestations of the genotypic alleles there behind. In humans, our keenest interest is to ask: On which chromosome is a disease mutation (gene) located? But let's see how different the conditions are for genetic studies in humans, *versus* in fruit flies.

2.6.2 THE CHALLENGES THAT WE FACE IN HUMAN GENETICS AND THE APPROACHES WE TAKE

The challenges that we face in human genetics, usually not problems in model organism genetics, include the following:

- In single or small family mating events, the progeny number is very small, not permitting an accurate calculation of the recombination frequency.
- Family mating history is not always available, making it difficult to determine the recombinants.
- No test mating can be performed, having to work with the outcomes but not planned experiments.
- No artificial mutagenesis can be performed, having to rely on naturally occurring variants. These tended to be other physiologically manifested phenotypes, or diseases, in the early days of human genetics.

The approaches that we take to overcome these challenges, but not always giving us the solutions, include the following:

- Small progeny number: The documented world record of the number of children by the same couple was 69, with skepticism regarding its authenticity. The couple were Feodor Vassilyev of Russia and his first wife Valentina in the 1700s. Feodor had another 18 children with his second wife. The total number of children by Feodor was 87, still a bit smaller than the total number of progeny required in the garden pea experiments performed by Mendel, and far smaller than the numbers of luxury in the fruit fly experiments by Morgan and colleagues, at about 1,000. For Mendel's results to have the required statistical power by the stringency of standards today, the progeny numbers did not need to be very large as the null hypothesis was accepted—that there is no difference among the groups. In Morgan's fly experiments, the much higher progeny numbers were needed to reject the null hypothesis in the event of linkage. This was especially necessary if the two loci studied were not very close and the deviations from the 50% recombination frequency were small.

  The plain truth but not always transparent to everyone including the geneticists is that the genes (loci) are universal in the same species. For the consideration of the locus–locus relationship between a disease locus D/d and a reference or marker locus M1/M2, all human mating events, whenever, wherever, between whomever are all eligible events. The only issue preventing the incorporation of all mating events into one "experiment" using not planned but existing outcomes is that the recombinants of the alleles across the two loci in the individual mating events cannot be lumped together.

  The situation can be compared to a game of coin tossing. If we intend to determine if a coin was minted one side heavier than the other, we can perform coin tosses to see if it lands on one side more frequently than on the other side. The fairest way to do this would be to use a well-calibrated mechanical device that tosses the coin consistently and do it 100 times. If the result is 70 times heads and 30 times tails, the coin is heavier on one side. If the result is about 50 times each side, the coin is even on both sides. But if the tossing is limited to 10 times, a 7 *versus* 3 result is not enough for us to draw any conclusion.

  Here we model the mold used to mint the coin as the loci under study in genetics. The coins that come off the mold as the individual persons carrying the alleles at the loci. In the practical setting of human genetics, different couples in different places (times) having 10 children *per* family would be for them to independently take a coin from the same minting mold, and toss it 10 times. The coin toss results individually can range from 10:0, 9:1, 8:2, 5:5 to 2:8, 1:9, 0:10. Is there still a way to analyze the data and draw a conclusion on the fairness of the coin (mold)?

Now let us consider the human genetics question: What is the chromosome relationship between the two loci, "Disease" with alleles D/d and "Marker" with alleles M1/M2? For different couples in different places (times) having 10 children *per* family, the recombinants *versus* nonrecombinants in the children can range from 10:0, 9:1, 8:2, 5:5 to 2:8, 1:9, 0:10. We cannot lump the recombinants *versus* the nonrecombinants from the different families and calculate the recombination frequency (fraction). The odds ratio for each family to present the particular result of recombinants and nonrecombinants by linkage or random assortments can, however, be calculated. The odds ratio can then be combined to inform on linkage or a lack thereof.

- Family history not always available and no test mating possible: Whenever possible, large and multigeneration families manifesting a genetic disorder would be preferentially included in the studies. Consanguineous families are uniquely informative to human genetic studies because of the increased allele sharing.

- Very limited number of variants (phenotypically), with many impeding on reproduction or even survival: This was a critical difference between fruit fly genetics and human genetics. The localization of the HGD (*HGO*) gene to human chromosome 3 in the early 1990s, still the formative years of the molecular genetic sciences, took advantage of the location of the FHH gene, wherein mutations manifest as another genetic disorder. This was quite literally the same as fruit fly genetics in practice, using two phenotypically manifested loci. It did not involve the calculation of recombination frequency (fraction), but involved "homozygosity mapping," that is, examining allele sharing in consanguineous families manifesting both diseases.

  It is how the FHH gene was localized to human chromosome 3 that told the story of how this challenge was met, and overcome in the most extensive and probably the most expensive human endeavor in biological science, known as the Human Genome Project. This endeavor gave us the chromosome differences at the molecular level, that is, DNA sequence. We obtained genetic markers that are polymorphic (more heterozygosity) and not associated with genetic disorders.

2.6.3 MAPPING OF HUMAN DISEASE GENES

In the end, the human genetic pursuit can be simplified and reduced to "gene mapping" by "linkage analysis." In principle, it is done like the proverbial keys on a chain and a lock. The keys are tried one at a time, until the one that opens the lock. The "Disease" locus with D/d alleles is the lock, and the "Marker" locus with M1/M2 alleles is the first key to test.

Step 1: Large multigeneration families manifesting the disorder are recruited and enrolled in the study.

Step 2: The genotype of the disease locus of each family member is determined as much as possible via analysis of the mode of inheritance (**Figure 2-4**).

Step 3: The genotype of the marker locus of each family member is determined using molecular biology techniques.

Step 4: The recombinants in the children are determined, when the linkage phase can be deduced, that is, from which grandparents the alleles at the two loci in a double-heterozygote parent are inherited, for example, D-M1 from one grandparent and d-M2 from the other.

Step 5: The odds ratio for each family to present the recombinants and the nonrecombinants by linkage *versus* random assortment is calculated. The data are combined to inform on linkage or a lack thereof.

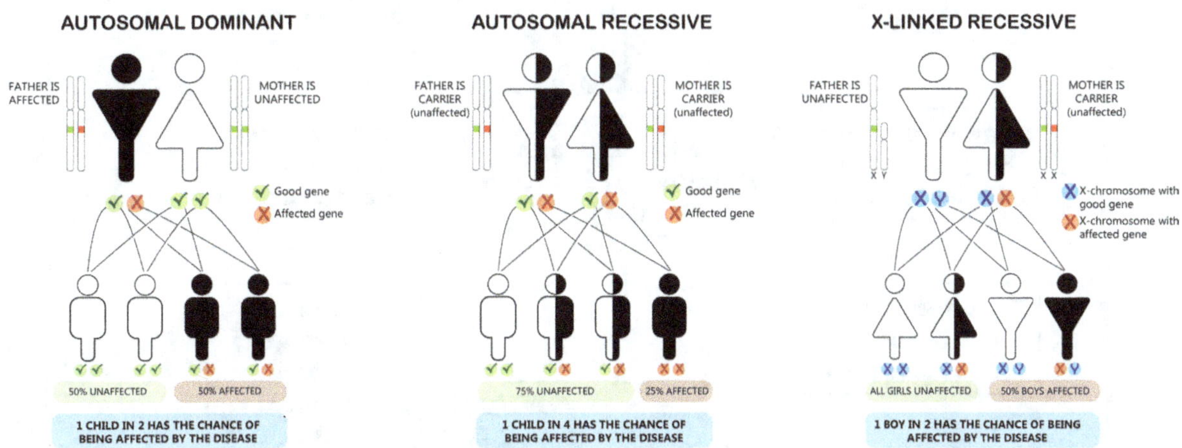

Katrien Francois/Shutterstock.com

FIGURE 2-4. Ascertaining mode of inheritance in human genetic diseases

If there is evidence for linkage over random assortment, the first key opened the lock! If the evidence is otherwise, a second marker is employed and the steps above are repeated, as necessary for additional markers, until one that shows linkage with the disease locus. Markers that are densely located around the first linked marker will then be employed to zoom in on the disease gene. The variants that cause the diseases could be identified by interrogating the DNA sequence at the final marker loci using molecular biology techniques. Otherwise, additional steps need to be taken to identify a candidate functional gene, for example, the steps that were taken to get to the human HGD (*HGO*) gene. **Figure 2-5** gives a flowchart view of the reverse genetics paradigm applied in the mapping of human disease genes by linkage analysis. In **Chapter 7**, we will go into details as to how an RF is determined in human genetic studies.

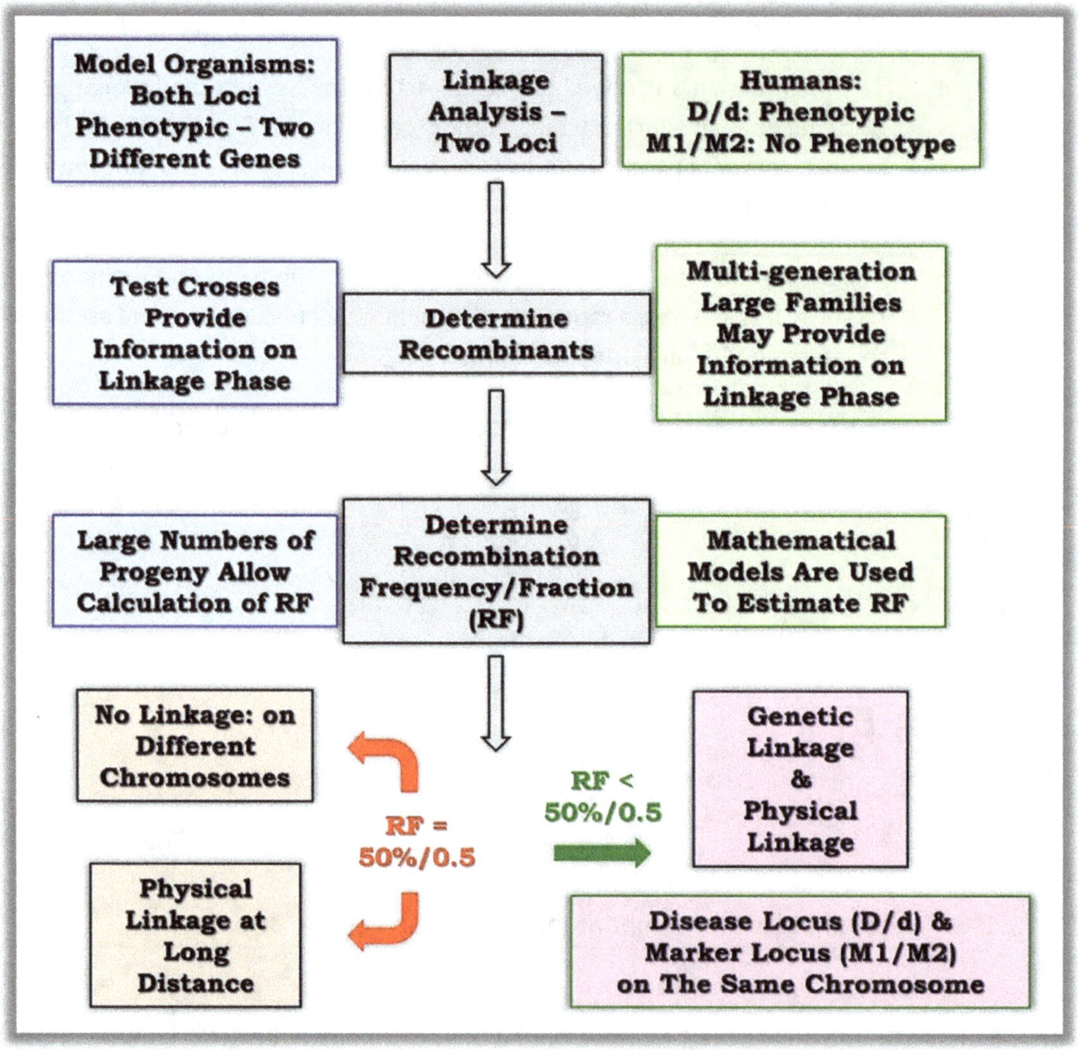

FIGURE 2-5. Reverse genetics in mapping human disease genes by linkage analysis, and the differences between model organism genetics and human genetics in practice

Now we return to **Table 2-1**, and look at the other rows of data, "phenotype description, molecular basis known," "phenotype description or locus, molecular basis unknown," and "other, mainly phenotypes with suspected Mendelian basis," with 6,425, 1,388, and 1,633 entries, respectively; and a total number of over 9,000. An example for "phenotype description, molecular basis known" is #104300, Alzheimer disease, familial, 1; AD1, with the β-amyloid precursor protein (APP) gene on chromosome 21q21 as a candidate susceptibility gene. For "phenotype description or locus, molecular basis unknown," an example is %102 300, restless leg syndrome, Susceptibility to, 1; RLS1, which is described to have been mapped to chromosome 12q. In "other, mainly phenotypes with suspected Mendelian basis," we have 231630, monosodium glutamate sensitivity, aka, Chinese restaurant syndrome. These entries reflect the full range of the unknown in human genetics and continue to be updated with new findings. Genetics, especially when it is applied to humans, is not an ancient science. We will look at how it is practiced today to solve the remaining and growing human genetic mysteries in the ensuing chapters.

BIBLIOGRAPHY

1. https://www.genome.gov/10001356/june-2000-white-house-event
2. https://www.omim.org/help/about
3. https://www.niaaa.nih.gov/alcohols-effects-health/alcohols-effects-body
4. https://www.niaaa.nih.gov/publications/alcohol-metabolism
5. https://www.fda.gov/news-events/press-announcements/fda-approves-first-gene-therapies-treat-patients-sickle-cell-disease
6. https://www.worthington-biochem.com/products/trypsin
7. Andersen DH. The present diagnosis and therapy of cystic fibrosis of the pancreas. *Proceedings of the Royal Society of Medicine*. 1949 Jan;42(1):25–32.
8. Bhattacharyya KB. The story of George Huntington and his disease. *Annals of Indian Academy of Neurology*. 2016 Jan–Mar;19(1):25–8.
9. Boyd Y, Buckle V, Holt S, Munro E, Hunter D, Craig I. Muscular dystrophy in girls with X; autosome translocations. *Journal of Medical Genetics*. 1986 Dec;23(6):484–90.
10. Bridges CB. Non-disjunction as proof of the chromosome theory of heredity. *Genetics*. 1916 Jan;1(1):1–52.
11. Bridges CB. Current maps of the location of the mutant genes of *Drosophila melanogaster*. *Proceedings of the National Academy of Sciences of the United States of America*. 1921 Apr;7(4):127–32.
12. Bridges CB. Salivary chromosome maps. With a key to the banding of the chromosomes of *Drosophila melanogaster*. *Journal of Heredity*. 1935;26:60–4.
13. Chou YH, Brown EM, Levi T, Crowe G, Atkinson AB, Arnqvist HJ, Toss G, Fuleihan GE, Seidman JG, Seidman CE. The gene responsible for familial hypocalciuric hypercalcemia maps to chromosome 3q in four unrelated families. *Nature Genetics*. 1992 Jul;1(4):295–300.
14. Ehlers MR, Fox EA, Strydom DJ, Riordan JF. Molecular cloning of human testicular angiotensin-converting enzyme: the testis isozyme is identical to the C-terminal half of endothelial angiotensin-converting enzyme. *Proceedings of the National Academy of Sciences of the United States of America*. 1989 Oct;86(20):7741–5.
15. Edenberg HJ. The genetics of alcohol metabolism: role of alcohol dehydrogenase and aldehyde dehydrogenase variants. *Alcohol Research & Health*. 2007;30(1):5–13.
16. Edenberg HJ, McClintick JN. Alcohol dehydrogenases, aldehyde dehydrogenases, and alcohol use disorders: A critical review. *Alcohol, Clinical and Experimental Research*. 2018 Dec;42(12):2281–2297.
17. Fernández-Cañón JM, Peñalva MA. Molecular characterization of a gene encoding a homogentisate dioxygenase from *Aspergillus nidulans* and identification of its human and plant homologues. *Journal of Biological Chemistry*. 1995 Sep 8;270(36):21199–205.

18. Fernández-Cañón JM, Granadino B, Beltrán-Valero de Bernabé D, Renedo M, Fernández-Ruiz E, Peñalva MA, Rodríguez de Córdoba S. The molecular basis of alkaptonuria. *Nature Genetics*. 1996 Sep;14(1):19–24.
19. Garrod AE. The incidence of alkaptonuria: A study in chemical individuality. 1902. *Molecular Medicine*. 1996 May;2(3):274–82.
20. Haldane JBS. The combination of linkage values, and the calculation of distances between the loci of linked factors. *Journal of Genetics*. 1919; 8: 299–309.
21. Hempel J, Holmquist B, Fleetwood L, Kaiser R, Barros-Söderling J, Bühler R, Vallee BL, Jörnvall H. Structural relationships among class I isozymes of human liver alcohol dehydrogenase. *Biochemistry*. 1985 Sep 24;24(20):5303–7.
22. Hoffman EP. The discovery of dystrophin, the protein product of the Duchenne muscular dystrophy gene. *FEBS Journal*. 2020 Sep;287(18):3879–87.
23. Hsu LC, Tani K, Fujiyoshi T, Kurachi K, Yoshida A. Cloning of cDNAs for human aldehyde dehydrogenases 1 and 2. *Proceedings of the National Academy of Sciences of the United States of America*. 1985 Jun;82(11):3771–5.
24. Ikuta T, Fujiyoshi T, Kurachi K, Yoshida A. Molecular cloning of a full-length cDNA for human alcohol dehydrogenase. *Proceedings of the National Academy of Sciences of the United States of America*. 1985 May;82(9):2703–7.
25. Ikuta T, Szeto S, Yoshida A. Three human alcohol dehydrogenase subunits: cDNA structure and molecular and evolutionary divergence. *Proceedings of the National Academy of Sciences of the United States of America*. 1986 Feb;83(3):634–8.
26. Jackson RE, Bellamy, MC. Antihypertensive drugs. *BJA Education*. 2015;15(6):280–5.
27. Jones RL, McDermott K, Drexler A, Niquette WS, Peterson CM. The role of insulin purification: An overview. *Diabetes Care*. 1982 May–Jun;5 Suppl 1:102–3.
28. Kannan TP, Zilfalil BA. Cytogenetics: past, present and future. *Malaysian Journal of Medical Sciences*. 2009 Apr;16(2):4–9.
29. Karas JA, Wade JD, Hossain MA. The chemical synthesis of insulin: An enduring challenge. *Chemical Reviews*. 2021 Apr 28;121(8):4531–60.
30. Kaufman TC. A short history and description of *Drosophila melanogaster* classical genetics: Chromosome aberrations, forward genetic screens, and the nature of mutations. *Genetics*. 2017 Jun;206(2):665–89.
31. La Du BN, Zannoni VG, Laster L, Seegmiller JE. The nature of the defect in tyrosine metabolism in alcaptonuria. *Journal of Biological Chemistry*. 1958 Jan;230(1):251–60.
32. Li H, Gu S, Cai X, Speed WC, Pakstis AJ, Golub EI, Kidd JR, Kidd KK. Ethnic related selection for an ADH Class I variant within East Asia. *PLoS One*. 2008 Apr 2;3(4):e1881.
33. Nurk S, Koren S, Rhie A, Rautiainen M, et al. The complete sequence of a human genome. *Science*. 2022 Apr;376(6588):44–53.

34. Pollak MR, Chou YH, Cerda JJ, Steinmann B, La Du BN, Seidman JG, Seidman CE. Homozygosity mapping of the gene for alkaptonuria to chromosome 3q2. *Nature Genetics.* 1993 Oct;5(2):201–4.
35. Riggs AD. Making, cloning, and the expression of human insulin genes in bacteria: The path to Humulin. *Endocrine Reviews.* 2021 May 25;42(3):374–80.
36. Roberts DB. *Drosophila melanogaster*: The model organism. *Entomologia Experimentalis et Applicata.* 2006;121:93–103.
37. Rubin GM, Lewis EB. A brief history of Drosophila's contributions to genome research. *Science.* 2000 Mar 24;287(5461):2216–8.
38. Schmidt SR, Müller CR, Kress W. Murine liver homogentisate 1,2-dioxygenase. Purification to homogeneity and novel biochemical properties. *European Journal of Biochemistry.* 1995 Mar 1;228(2):425–30.
39. Soubrier F, Alhenc-Gelas F, Hubert C, Allegrini J, John M, Tregear G, Corvol P. Two putative active centers in human angiotensin I-converting enzyme revealed by molecular cloning. *Proceedings of the National Academy of Sciences of the United States of America.* 1988 Dec;85(24):9386–90.
40. Sturtevant AH. The linear arrangement of six sex-linked factors in Drosophila, as shown by their mode of association. *Journal of Experimental Zoology.* 1913; 14:43–59.
41. Sures I, Goeddel DV, Gray A, Ullrich A. Nucleotide sequence of human preproinsulin complementary DNA. *Science.* 1980 Apr 4;208(4439):57–9.
42. Takemori S, Furuya E, Mihara K, Katagiri M. Bovine liver homogentisicase: Apoand reconstituted holoenzymes. *European Journal of Biochemistry.* 1968 Nov;6(3):411–8.
43. Tanimoto K, Sugiyama F, Goto Y, Ishida J, Takimoto E, Yagami K, Fukamizu A, Murakami K. Angiotensinogen-deficient mice with hypotension. *Journal of Biological Chemistry.* 1994 Dec 16;269(50):31334–7.
44. von Bahr-Lindström H, Höög JO, Hedén LO, Kaiser R, Fleetwood L, Larsson K, Lake M, Holmquist B, Holmgren A, Hempel J, et al. cDNA and protein structure for the alpha subunit of human liver alcohol dehydrogenase. *Biochemistry.* 1986 May 6;25(9):2465–70.

3

DNA

The textbook of life but someone has to read and build

3.1 The Keepers, Copiers, and Readers of the Book of Life

The word DNA, might as well be a word, now, is in everyone's vocabulary. When Liverpool Football Club comes back from a losing position to win a match with little to no time remaining, the fans celebrate with the motto "'Never Give Up' is in our DNA." DNA stands for deoxyribonucleic acid, and it is the major form of what is called the "genetic material," the material onto which the book of life is printed.

The book of life is important for all of our body's cells, originating from the single fertilized egg, in which one copy is from the father and the other from the mother, at three billion base pairs *per* copy. An adult human has trillions of cells on average, 100 trillion by the popular estimate but 37 trillion by a published study. Every person, however, starts with just one cell. The single fertilized egg undergoes a division to become two, and with further divisions, the cell numbers will continue to double. The blastomere cells in the early cleavages are not identical at the molecular level, even at the two-cell stage, and the two cells contribute disproportionally to the body's cells. Further cell divisions will activate the ensuing molecular and cellular programs of embryonic development, leading to cells that are molecularly, functionally, and anatomically different in the developing and the fully developed body. At the anatomical and histological levels, there are about 200 types of cells in an adult human.

For each cell type that forms a part of a tissue architecture, cell division must generate two identical daughter cells. The genomic DNA of the two daughter cells is, for all intents and purposes, an exact copy of the genomic DNA in the no longer existing parent cell. There is a mutation rate in the DNA replication process, but at a rather low level. The cell's "mismatch repair system" removes most of the errors in the time window before the next round of replication, and additional repair systems take care of chemical damages. Tissue-specific mutations in humans range from 9 *per* year in the bile ductular cells to 56 *per* year in the appendiceal crypts. The molecular differences, at the messenger RNA (mRNA) and the protein levels, in the various bodily cells, called the somatic cells, are the results of restricted and selective use of the genetic information in the genomic DNA. The cell's DNA is not wide open for access by the molecular machinery for transcription, making a (complementary) copy of the gene in the form of a messenger RNA (aptly named) to send to the protein-making machinery outside the nucleus for production. The mRNA is the blueprint of the original technical engineering drawing, the DNA. If everything encoded in the DNA is made by every cell in the body all the time, every cell in the body at any time would be the same as the next and they can only perform one set of functions, without the possibility of changes or the ability to adapt to the changes in the cell's environment.

The relatively new topic of epigenetics is already a quarter century old, now. The basic idea is that the genomic DNA in our cells is restricted to transcription by the postsynthesis modifications of the DNA itself, and the chromosome proteins that act to regulate transcription. For example, the DNA can be methylated, that is, the addition of a methyl ($-CH_3$) group to the nucleobase cytosine at the 5-position. The DNA in the nucleus is never naked and is always bound to the core histone proteins, which have their amino-terminal protein tails wrapped around the DNA. During DNA replication and transcription, the polymerases and the auxiliary proteins will have to negotiate around the core histone proteins, with assistance by additional protein factors. The histone tails can be methylated or acetylated, rendering the wrapping of DNA tighter or more relaxed. The first word used in history to describe the chromosome was "chromatin," which was soon retired for that purpose. It is now being used to describe the finer molecular details of the chromosome structure, especially in the context of whether the structure is in an open or closed state to allow or restrict the access of DNA to the transcription machinery.

There is coordination between the DNA methylation and the histone modifications through the actions of other protein factors. These are the simplest examples of how restriction and access to the genomic DNA for transcription can be regulated. Throughout the personal development of a human, these modifications are dynamic and change in response to the environment, down to the cellular level. The established cell identity within a tissue is maintained by copying over the DNA and the histone protein modification marks during chromosome duplication before division and distribution into the daughter cells.

All biochemical reactions in the body obey the chemical reaction laws, which can be condensed into two. First, every reaction is passive, and second, every reaction is concentration driven, that is, by the quantities of the reactants and the products. The genetic information contained in the DNA, either the beautifully modeled double-stranded (ds) form or the single-stranded (ss) form, is decoded by other molecules. It is easier and more comfortable for us to use a personified language to describe the biological processes, such as "a transcription factor recognizes a specific DNA sequence in the promoter/enhancer region of the gene to regulate the gene transcription." The reality is that the recognition is the formation of a stable complex between the transcription factor protein molecule with its unique three-dimensional structure and the major groove of the dsDNA with that specific sequence.

The information contained in the DNA is silent. The physical appearance of the purified dry form of high molecular weight, that is, very much intact DNA, from animal cells, is white and stringy, and totally lifeless. The Swiss physician Friedrich Miescher who was the first to isolate DNA, initially named nuclein, from the pus on wound bandages and fish sperm in the 1860s to 1870s, dismissed it as a candidate substance to transmit hereditary traits. His reason for doing so was that the form of DNA seemed very consistent and lacking in diversity from the different sources including the reproductive cells such as the fish sperm. It would not be able to fulfill the range of molecular functions in the cells that make up an animal just by itself. This is still true, today.

3.2 The DNA Double Helix Has the Perfectly Identical Base-Pairing Geometry

The double helix of DNA is no longer a structurally technical concept of a biochemical molecule, but a symbol of biological or medical science for everyone in all walks of life, appearing in the logos of countless companies. We will not review the history behind the discovery of this beautiful and symbolic structure but highlight the most relevant features of it to help make "sense" of biology and genetics.

The double helix is formed by two anti-parallel strands of DNA (**Figure 3-1**). Each strand is a chain of nucleosides consisting of a deoxyribose and a nucleobase, linked by a phosphodiester bond. The nucleobases in the DNA are adenine (A), guanine (G), cytosine (C), and thymine (T), that is, the letters of the genetic information language. Adenine and guanine each contains nine atoms and are called purines. Cytosine and thymine each contains six atoms and are called pyrimidines. The atoms in the nucleobases are numbered 1 to 6 for pyrimidines and 1 to 9 for purines. The deoxyribose is a five-carbon sugar ring, with the carbons numbered by 1′ to 5′. The biochemical reaction by which the DNA chain is formed relies on the 3'-OH (hydroxyl) group to connect to the α-phosphate attached to the 5′-carbon of an incoming nucleoside triphosphate. Thus, the DNA strand has a chemically determined direction, in the same direction as its biochemical synthesis, 5′ → 3′. The two strands in a double helix run in opposite directions, hence anti-parallel. The specific A-T and G-C base pairing gives an identical geometry and is the most important feature of the double-helical DNA.

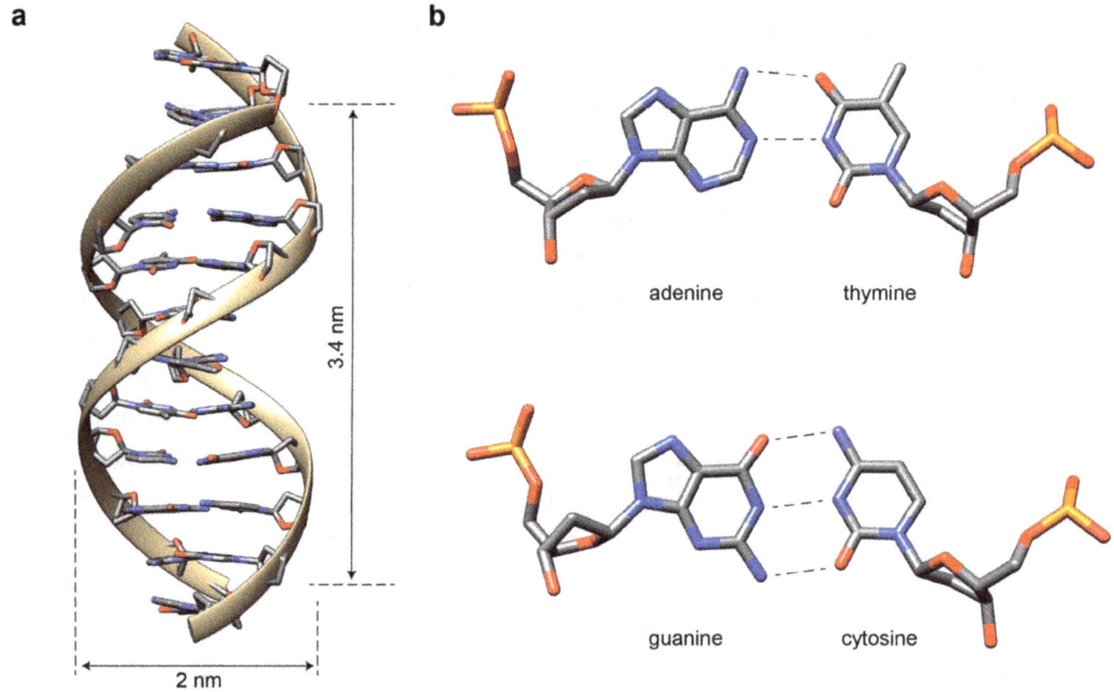

FIGURE 3-1. The structure of double-stranded DNA and the identical single geometry of the A-T and G-C base pairs

The bases attached to each strand in the double helix are internal in the structure, and face each other on the opposite strands, in pairs. Each nucleotide with the phosphate *per* unit is the same in size; thus, the two bases in a pair are level to each other. For naturally occurring dsDNA, the base pairs are A to T or G to C, a nine-atom ring to a six-atom ring. The forces that pull the paired bases together are called hydrogen bonds, formed between a proton (hydrogen) donor and a proton acceptor on the opposite nucleobase rings. There are two hydrogen bonds between the A-T base pair and three between the G-C pair. The significance or impact of an individual hydrogen bond or even the multiple between an isolated base pair is negligible. Hydrogen bonds become a real force of structural stabilization when a large number can form together, such as those on the inside of the double-helical DNA of a good length. The best example of understanding the collective power of hydrogen bonds is water. Two water molecules can form a hydrogen bond, but it makes no sense to think of two water molecules existing in a pair. A cup of water has a molar concentration of 55.6, and the number of hydrogen bonds becomes massive to allow us to pour water by its cohesive property. Yet, the water molecules on the surface in an open cup would still evaporate because the collective hydrogen bonds at the surface are not strong enough to hold down the water molecules.

3.3 DNA Replication—Making Precise Copies of DNA for Transmission

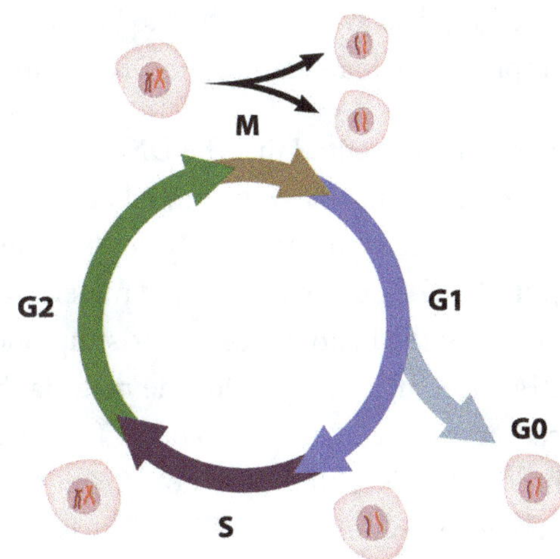

Alila Medical Media/Shutterstock.com

FIGURE 3-2. The cell cycle is marked by different phases of molecular and cellular events.

Theoretically, on a one out of many billion (or much more) chances, the formation of the 3′,5′-phosphodiester bonds between the nucleotides in a chain of DNA can happen unassisted. It would require the 3′-OH group on the existing DNA chain to point at the incoming deoxyribonucleoside triphosphate (dNTP, N for any kind) on the α-phosphate at a precise angle and distance. This would be as easy as pulling two water vapor molecules out of the air and bonding them together. Clearly, any form of life with DNA as the genetic information keeper cannot rely on a mechanism like such for replication. On average, a human cell divides every 24 hours, and all molecular and cellular business needs to be finished and ready for the division in this short time window. A complete program of cell division is called a cell cycle, divided into four phases, each with characteristic and critical biochemical and cellular events (**Figure 3-2**).

3.3.1 THE CELLULAR PRECISION OF DNA REPLICATION

The DNA is replicated in the S (synthesis) phase, about 8 hours long on average for human cells. But before the synthesis can begin, a cell mobilized to divide and enter the cell cycle needs to prepare the right types of materials in the right amounts. The material preparation occurs in the G1 (gap 1) phase, about 11 hours long on average for human cells. Between G1 and S, there is a "checkpoint" at the junction, to ensure that the material preparation is complete. The so-called checkpoint is quite unlike human checkpoints where proper papers are examined before passage is allowed. The G1-S junction checkpoint function is fulfilled at the molecular level by the passive laws of chemical reactions. We will look at a couple of really good and telling examples of how this works.

First, to prevent the duplicated chromosome arms from totally tangled up, as 46 will become 92, the sister chromatids need to be locked up, handcuffed, to be exact. The protein complex that performs this function is called a cohesin, and a model of its structure shows a linked pair of handcuffs, each capable of locking a 30-nm DNA fiber, that is, DNA wrapped around the core histones and wrapped up by the core histone tails and further compacted by the binding of the linker histone H1. It would make no sense to do this after 46 becomes 92, and it would make the perfect sense that the cohesin complexes are loaded onto the DNA before replication.

Second, it is essential that the DNA is not left naked over a long stretch or a long stretch of time, the nascent DNA needs to be immediately wrapped onto the core histones and up by the histone tails. The DNA needs to be precisely doubled before and after the S phase, no less but also no more. Intuitively, for the sake of energy and material conservation, the cells would only need to make one extra portion of the histones, as the existing ones are kept and evenly distributed onto the two daughter DNA molecules, not just for the material but more importantly for inheriting the epigenetic modification marks. With a ratio of 1:1 for the new DNA and histones, however, the process of ensuring immediate DNA binding is not efficient by the laws of chemical reactions, which are always passive and concentration driven. The fact is histones are made hundreds of fold extra than the one-fold extra required during the S phase. The mobilization of histone production actually starts in the G1 phase. The S phase does not begin until the cellular molecular programs, via passive quantitative chemical reaction laws, sense the quantitative accumulation of histones and the loading of the cohesin complexes onto the prereplication DNA.

3.3.2 THE MOLECULAR PRECISION OF DNA REPLICATION

The human genome is three billion base pairs long and each cell has two copies, so six billion base pairs of DNA need to be precisely replicated in the S phase. The polymerization reaction cannot be left at the level of random collision chemistry for the 3'-OH to attack the α-phosphate of an incoming dNTP by chance. DNA polymerization happens in the cavity of a DNA polymerase, which serves as a molecular sieve. The dsDNA is unwound and opened in the region of active replication as single-stranded templates. The immediacy of the DNA replication makes it more advantageous to start the process with a short RNA primer, because inside the nucleus the concentrations of the ribonucleoside triphosphates (NTPs) are much higher than that of the dNTPs, at between 10- and >100-fold. The RNA primer is later polished away and replaced with DNA by specialized enzymes.

The DNA polymerase has a high affinity toward the primer-template junction. Upon binding, the template strand passes through the cavity of the DNA polymerase, wherein only the A-T or G-C pairs can fit due to the constraint of the single geometry. Thus, opposite an A on the template strand, only a dTTP can be accommodated so accurately to align its α-phosphate with the 3'-OH on the nascent DNA or the RNA primer. The engagement of a DNA polymerase with the primer-template complex takes about 1 second, and the enzyme has a tendency to fall off after a short stretch of elongating the existing DNA with new nucleotides. A specialized ringed protein complex called a sliding clamp is employed to keep the polymerase engaged with much improved "processivity," that is, synthesizing a long stretch of DNA after binding to the primer-template junction without falling off.

The two strands in the original (parent) DNA are complementary to each other, while the DNA polymerase selects the nucleotides in the newly synthesized DNA by base-pairing complementation, too. This reciprocal complementary relationship between the two template strands and the two new strands means that the two new dsDNA, each with one original strand and one new strand, are identical. Thus, no further sorting of original *versus* new strands is required at the end of the DNA replication process. This was the hypothesis of "semiconservative" DNA replication tested and proven by Matthew Meselson and Frank Stahl in 1957 in their classic experiment using the *Escherichia coli* DNA. The first DNA polymerase was isolated in 1958 from *E. coli*, by Arthur Kornberg (1918–2007).

3.4 Transcription—Making "Sense" of the Genetic Information Stored in DNA

The decoding of the genetic information, the residue-by-residue nucleotide sequence, is called transcription, in quite literally the precise sense of the word. The information is now "transcribed" into a different molecular carrier (onto a different sheet of paper), the RNA. The biochemical reaction is similar to that in the DNA chain elongation, a nucleophilic attack of an incoming NTP on the α-phosphate by the 3'-OH on the existing end of an RNA, or on the first nucleotide at the initiation of transcription. Transcription initiation is regulated by many protein factors and by the accessibility of DNA. The "open or closed" state of the chromatin structure in a chromosome region for transcription is controlled by chemical modifications of the DNA and the histone tails that wrap around the DNA. The RNA polymerases also operate in a similar manner as the DNA polymerases, selecting the incoming NTP for incorporation into the nascent strand by base pairing to the DNA base in the first open position in the template strand. A uracil (U) in the NTPs pairs with an A in the template DNA and is incorporated into the RNA. The early work on protein-coding gene transcription by RNA polymerase II in eukaryotic cells was pioneered by Roger Kornberg, son of Arthur. Both the father and the son received a Nobel Prize for their contributions to our understanding of molecular biology.

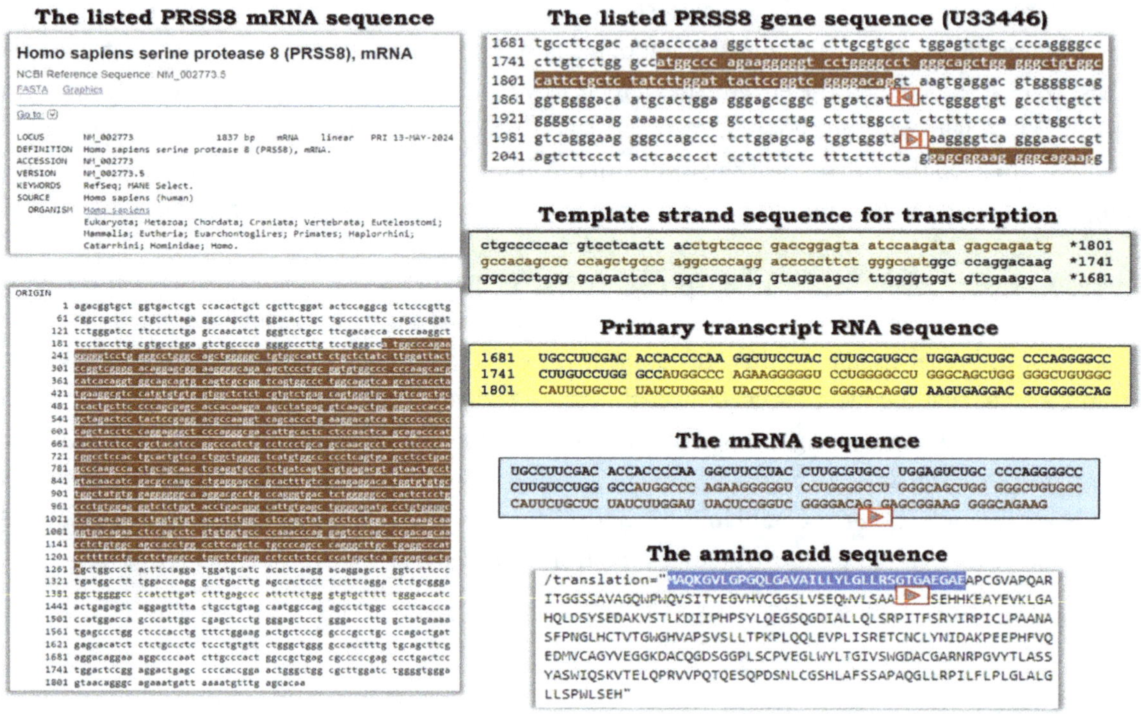

FIGURE 3-3. Making "sense" of the genetic information stored in DNA

The mRNA sequence of the serine protease prostasin (PRSS8), under the accession number NM_002773, is listed with "t" in the sequence instead of "u." The PRSS8 gene sequence, under the accession number U33446, lists the sense strand, which corresponds to the mRNA sequence in the coding regions, highlighted in brown. The template strand of the PRSS8 gene is a reverse complement of the listed sense strand. The asterisked numbers correspond to the numbers in the sense strand, but these are not the numbers for the template strand. In the primary transcript, cap letters and "U" are used. The mRNA sequence is a partial sequence containing the spliced first two exons. The exon-intron junctions are indicated by the "advance to stop" buttons, and the spliced junction in the mRNA and the corresponding junction in the protein sequence are indicated by the "play" button.

In human cells, the protein-coding genes are transcribed to produce a complete complementary copy, the primary transcript, which is then processed to produce the mRNA. The mRNA is shipped from the nucleus to the cytosol, and the genetic information transcribed over from the DNA is decoded by the anticodon of transfer RNAs, by base pairing with the triplet codons in the mRNA. The amino acids carried by the tRNAs are then assembled in the order of the triplet codon sequence in the mRNA to produce a polypeptide. A properly folded polypeptide into a three-dimensional structure with biochemical activities is now a protein, executing life's essential functions. In this sense, the sense in the biological information is defined at the level of the amino acid sequence in the protein. Thus, it would follow that the mRNA must be the sense strand. Because of base complementation in nucleic acid biosynthesis, dictated by the single and identical geometry of the A-T(U) and G-C base pairing, the strand in the dsDNA that is identical in sequence to the mRNA in the coding regions, except for T's in the positions of U's, is the sense strand. This strand is always listed in publications and databases as the corresponding "gene sequence," as shown in **Figure 3-3**. But in reality, it cannot be, as transcription from this strand would give you the antisense RNA, dead ending the central dogma. The strand in the dsDNA that is transcribed, the true gene sequence, is the one that is reversely complementary to the mRNA in the coding regions, the antisense DNA strand. Listing the sense strand of the DNA as the gene sequence is for the purpose of allowing us to intuitively relate to the mRNA sequence. For the same reason, the mRNA sequences in GenBank uses "t" for "u," giving an even more direct relationship between the two carriers of the same genetic information. The reason for the lowercase letters is to avoid potential visual confusion between the cap letters C and G.

3.5 The Genetic Information in Human DNA

3.5.1 THE MAKEUP OF OUR DNA

The starting point of the investigation into the text information contained in DNA was the ground-laying biochemistry work by Erwin Chargaff (1905–2002), who degraded (hydrolyzed) naturally occurring DNA from various sources, including humans, for analysis of the nucleotide composition, that is, how much of each kind is present in each DNA sample. In the early 1950s, Chargaff presented data to show that in these DNA samples, the amount of A nucleotides is the same as that of the T nucleotides, and the amounts of G and C nucleotides are the same, as well. This finding predates the famous double-helix model of DNA of Watson and Crick (1953) and was a key piece of the final puzzle for the resolution of the DNA structure.

Chargaff's intent was to test out the tetranucleotide hypothesis of Phoebus Levene (1869–1940) that DNA is composed of repeating sequences of the four nucleotides. Under this hypothesis, it would follow that the four nucleotides have an equal abundance, or with a composition of 25% each. It would also dictate that all species have an identical nucleotide composition of their DNA. Prior to analyzing the human DNA nucleotide composition, Chargaff discovered that the compositions of DNA from different sources were entirely different, these included calf/ox, yeast, and the bacterium avian tubercle bacilli. DNA from different tissues of the same species, such as calf thymus and ox spleen, however, have the same composition. The tetranucleotide hypothesis was completely disproven by Chargaff's work. The idea of DNA containing repeating nucleotides is, however, not only not lost but becoming quite important and directly relevant to the study of genetics and molecular medicine.

In humans, the proportions of the nucleotides are, on average, 30% A and T, and 20% G and C, not very different from the reported proportions in Chargaff's analysis of human spleen and thymus DNA in 1950. By the common estimate, 50% of the human DNA are composed of repetitive sequences, while the genes are contained in the nonrepetitive half. Protein-coding genes occupy about 41% of the genome, but these include the entire "structural genes" by the modern molecular descriptions, such as the regulatory regions where RNA polymerase II and the transcription factors would bind. The actual coding sequences, that is, the exon sequences, comprise only less than 2% of the human genome. Annotation of the human genome is highly challenging, and the rules and methods are ever-changing and evolving with our improved understanding and the revolution in technology. For example, 80.4% of the genome are associated with some kind of biochemical activity, potentially manifesting as a phenotype. At present, the commonly accepted number of the protein-coding genes in the human genome is around 20,000.

3.5.2 HOW AND WHERE ARE WE GENETICALLY DIFFERENT?

For our purposes in the application of human genetic studies, we are more keenly interested in how and where in the genome individuals can be different. The Human Genome Project provided for us a Reference Human Genome, a best possible consensus representation of the different human populations. When compared to this reference genome, an individual person's genome of six billion base pairs may have about 5,000,000 single-nucleotide variants (SNVs), about 600,000 insertion/deletion variants totaling to 2,000,000 nucleotides, and about 25,000 structural variants totaling to 20,000,000 nucleotides. Overall, approximately 27 million base pairs, or 0.4% may be different in a person's genomic DNA sequence in various places from the reference genome. We infer from this that any two individuals' DNA would be 0.4% different, or 99.6% identical.

The single-nucleotide polymorphisms (SNPs): Of particular interest and utility to human genetic studies are two types of variants, known as single-nucleotide polymorphisms (SNPs) and microsatellites (2–10 base-pair repeats). The SNVs with the minor allele at a frequency >1% in the population are classified as SNPs. Chemically mechanistically, the likelihood of any of the four DNA nucleobases to be represented at an SNP is the same for all four. The great majority of the SNPs in humans, 99.679%, are only choices of two of the four, or biallelic, as shown by the analysis of the data from the 1000 Genome Project. The reason behind this is the low mathematical probability for two independent germline changes to occur in the same nucleotide position and to survive and expand in the population. A mechanism was proposed to explain the triallelic SNPs with simultaneous changes on both strands of the DNA during DNA replication or by chemical damage, with an escape from the DNA repair programs before the next round of replication.

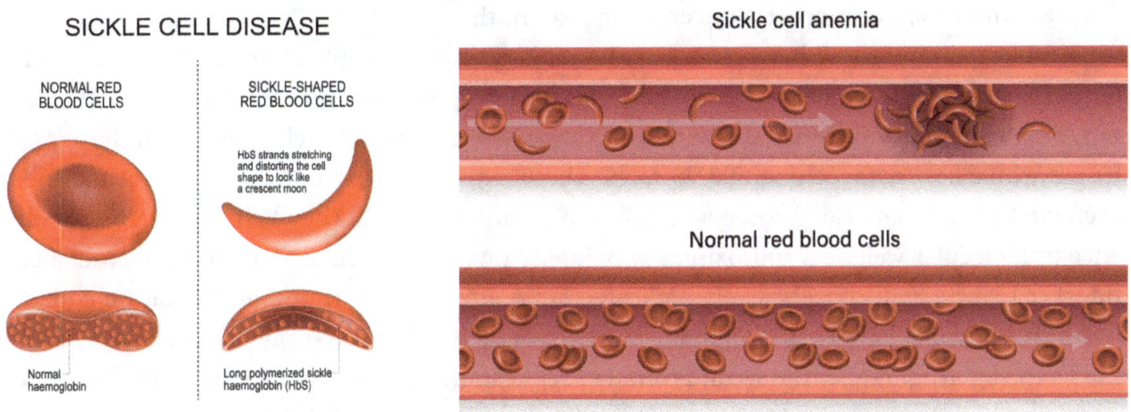

Designua/Shutterstock.com

Dee-sign/Shutterstock.com

FIGURE 3-4. Sickle cell disease (anemia)

An example of an SNP is rs334, where "rs" is the reference SNP (refSNP) cluster ID number. A refSNP cluster may contain one or several submitted SNPs. SNP rs334 is on human chromosome 11, at position 5,227,002 (GRCh38 38.1/141), in the coding region of the human β-globin gene. The normal allele "A" is the middle base in a Glu codon GAG, whereas the sickle allele "T" changes the codon to GTG, coding for Val. The substitution of Val for Glu at the sixth position of the mature β-globin polypeptide changes a charged side chain to a hydrophobic one, allowing the sickle cell β-globin (HbS) polypeptide to make a polymer in the absence of oxygen via hydrophobic interactions with the Phe in position 85 and Leu in position 88. The HbS polymers deform the red blood cell, that is, rendering it sickle in shape, resulting in a vaso-occlusive crisis in the capillary venous pole in the sickle cell disease (anemia) (SCD) (**Figure 3-4**).

The 1000 Genome Project analyzed some 84 million SNPs so clearly not all would be the basis of a genetic disorder manifested at the physiological or physical level phenotypically. As a matter of fact, most are not. With their distribution across the entire genome, we would have many loci to use as genetic markers. The mostly biallelic nature of the SNPs presents a limiting factor in their use as genetic markers, as the heterozygosity is limited by the choices of only two alleles.

The microsatellites: The microsatellite repeats are more advantageous as genetic markers. The namesake of satellite DNA came from the very early days in the molecular analysis of DNA from various sources. The DNA from bacteria formed a single band in the test tube after being spun into a dense salt (CsCl) gradient medium by ultracentrifugation, essentially a test of the molecule's buoyant density. Molecules stopped moving when they reached the point of the salt gradient medium with an identical density to theirs.

The best known application of ultracentrifugation in the analysis of DNA was the inspired Meselson-Stahl experiment. With the incorporation of N-14 isotope in the DNA of *E. coli* cells cultured in N-15 medium, Matthew Meselson and Frank Stahl provided the experimental evidence that DNA replication is semiconservative. The *E. coli DNA* is approximately five million base pairs in length with a base composition of 25% each. There are 15 nitrogen atoms for all four nucleobases, averaging to 3.75 *per* base. Thus, between a heavy (old) and a light (new) chain of DNA, the difference in molecular weight is approximately 3.75/327 (average molecular weight of the four nucleotides), 1.15%. A semiconservatively replicated DNA after the first round of replication containing one heavy and one light chain would be 0.575% from the old (heavy-heavy) and also 0.575% from the DNA (light-light) after the second round of replication. Meselson and Stahl were able to clearly and unambiguously observe the three distinct forms of *E. coli* DNA, lending proof to their hypothesis of semiconservative DNA replication.

The DNA from bacteria, relatively short in the range of several million base pairs, has a rather uniform density. Intuitively, the double-helical structure of DNA would suggest that physically or chemically, or physical-chemically, any and all segments of DNA have the same properties. The nucleotides do indeed have different molecular weights, and the base pairs differ slightly, too. The molecular weight of a G-C nucleotide pair is 654.4 and that of an A-T pair is 653.4. This level of difference, however, is far below the sensitivity of the CsCl gradient ultracentrifugation method.

But when DNA from some eukaryotic species, such as the crab, salmon, mouse, guinea pig, ox, and human, were analyzed by CsCl density gradient ultracentrifugation, two or more bands were formed. Quantitatively minor bands with higher or lower buoyant densities were named "satellite DNA," inspired by the Space Age in the 1960s, as if they were a satellite revolving around a planet—the major band. The bands can also be represented by peaks on the graphs of the DNA intensity scans up and down the centrifuge tubes containing the samples. The major band in human DNA has a CsCl buoyant density of 1.700 g/cm^3 and contains the bulk of the genomic DNA, while the minor band has a density of 1.687 g/cm^3 and contains less than 1% of the total DNA. The technology applicable to analyzing DNA content differences was limited in the 1960s to the early 1970s, and CsCl density gradient ultracentrifugation was the mainstay at the time. Today, we have many molecular biotechnology tools at our disposal so we would not think of using this method to investigate if the DNA extracted from a mouse tumor has gross changes as compared to the normal mouse DNA. At the time, however, this was one of the research goals.

The early results were limited to measuring out the differences in base composition between DNA samples from different species, bacteria *versus* bacteria, and animal *versus* animal. We might as well consider this the early infancy of forensic DNA technology. Along with these investigations, the art of using heat or chemical agents to separate and then reassociate the DNA strands was being perfected. From these studies, it was learned that the satellite bands contained DNA with repeating or repetitive sequences, as these would reassociate much faster than DNA of random or nonrepetitive sequences. It was also learned that the DNA samples in the satellite bands with the lower buoyant densities have more A-T base pairs, that is, AT rich, while the higher buoyant density satellite DNA samples are GC rich. The GC content and the buoyant density have an upward linear relationship. The differences in buoyant density between AT-rich *versus* GC-rich DNA are not caused by the differences in molecular weight (653.4 *versus* 654.4) or the number of hydrogen bonds in the base pairs (2 *versus* 3). In the CsCl solution, AT base pairs retain more water molecules than the GC base pairs. As a result, GC-rich DNA would be more compacted than AT-rich DNA, manifested as the differences in buoyant density.

3.5.3 THE REPETITIVE DNA IN HUMANS

The term *satellite DNA* was originally used only in reference of DNA separated by ultracentrifugation but has been broadened to include all repetitive DNA from a species. And in humans, this portion is at least 50% of the total DNA. There are several major categories of repetitive DNA, on the basis of the repeating sequence unit lengths in base pairs (bp) or the mechanisms of their origin. These include microsatellites (1–10 bp repeats), minisatellites (10–100 bp repeats), and small-scale repetitive elements (SSREs) (0.1–0.8 kbp repeats). Microsatellites are the most relevant to our discussion in the application to human genetics, and the size of the repeating units is rather arbitrarily set. These are also called simple sequence repeats (SSR) or short tandem repeats (STR), that is, head-to-tail direct repeats, as opposed to an inverted repeat. Micro- and minisatellites are collectively referred to as "variable number tandem repeats" (VNTRs). Currently, it is estimated that ~3% of the human genome are microsatellite DNA.

The most outstanding characteristic of a VNTR DNA is the number of alleles, differing in the number of repeats. Two major molecular mechanisms can lead to the VNTR length differences, replication slippage, and unequal crossing-over. Because of the repeating nature of the DNA in such regions, the replicative DNA polymerase can engage a primer-template junction with a loop of an ssDNA behind it, either on the template side or on the nascent strand side. This would appear to be that the DNA strands slipped a step or a few, as the engagement of the DNA polymerase with the primer-template junction is a dynamic process. The resumed replication would change the repeating number on the nascent DNA, a gain if the slippage is on this side, and a loss if the slippage is on the template side. The replicative error rate in the repetitive DNA is estimated to be 10- to 100-fold higher than that in the nonrepetitive and random sequence regions. The enzymes in the replication mismatch repair system are also less effective in the repetitive DNA region, especially for long stretches. In the next round of DNA replication, the changes in the repeating numbers become fixed. Replication slippage occurring in the pregametic cells, the oogonium and the spermatogonium, will lead to changed numbers of repeats in the mature egg or sperm, that is, creating new alleles. Also because of the repeating nature of the DNA in such regions, during meiosis division I, the pairing of the homologous chromosomes may be misaligned but still achieve perfect sequence matches to allow crossing-over. For the two participating non-sister chromatids, this crossing-over is unequal, that is, one becoming longer and the other shorter, creating new alleles in the gametes, too. For the rather short microsatellites, replication slippage is suggested to be the major mechanism.

The variable lengths of the microsatellite would offer the human genetic researchers the desired heterozygosity of the marker locus, much more than the biallelic SNPs can provide. It is also very unlikely that the repetitive DNA would land in the coding regions of genes. The very essence of protein function is the three-dimensional structure of the properly folded polypeptide relying on the different side chains in the amino acid residues, via intra- and interchain interactions. Thus, a long stretch of the same amino acid residue would be a major issue to the folding. A long stretch of microsatellite DNA would precisely dictate a long stretch of the same amino acid residue. Over the long course of adaptation of protein functions to serve the cellular and ultimately the organismal functions, the coding sequences of proteins tend not to have such features. But, there are exceptions, and one of such was actually the final key that opened the lock to the genetic puzzle of Huntington disease (HD).

The genetic effect of the microsatellites not in most protein-coding genes is that they are silent in the phenotypic selections during mating. So with the discovery, investigation, and characterization of the microsatellites, along with the majority of the SNPs, we had effectively overcome one major challenge in the practice of human genetics. We now have marker loci, that is, we know exactly where they are on the chromosomes (in the genome), and the majority of them are not associated with another genetic disorder.

BIBLIOGRAPHY

1. https://www.genome.gov/about-genomics/educational-resources/fact-sheets/human-genomic-variation
2. https://www.ncbi.nlm.nih.gov/snp/rs334
3. 1000 Genomes Project Consortium; Auton A, Brooks LD, Durbin RM, Garrison EP, Kang HM, Korbel JO, Marchini JL, McCarthy S, McVean GA, Abecasis GR. A global reference for human genetic variation. *Nature*. 2015 Oct 1;526(7571):68–74
4. Altemose N. A classical revival: Human satellite DNAs enter the genomics era. *Seminars in Cell & Developmental Biology*. 2022 Aug;128:2–14.
5. Bianconi E, Piovesan A, Facchin F, Beraudi A, et al. An estimation of the number of cells in the human body. *Annals of Human Biology*. 2013 Nov–Dec;40(6):463–71.
6. Burgers PMJ, Kunkel TA. Eukaryotic DNA replication fork. *Annual Review of Biochemistry*. 2017 Jun 20;86:417–38.
7. Chargaff E, Zamenhof S, Green C. Composition of human desoxypentose nucleic acid. *Nature*. 1950 May 13;165(4202):756–7.
8. Chargaff E. Some recent studies on the composition and structure of nucleic acids. *Journal of Cell Physiology* Suppl. 1951 Jul;38(Suppl. 1):41–59.
9. Chen Z, Li S, Subramaniam S, Shyy JY, Chien S. Epigenetic regulation: A new frontier for biomedical engineers. *Annual Review of Biomedical Engineering*. 2017 Jun 21;19:195–219.
10. Cobb M. 60 years ago, Francis Crick changed the logic of biology. *PLoS Biol*. 2017 Sep 18;15(9):e2003243.
11. Cooper GM. *The cell: A molecular approach*. 2nd ed. Sunderland, MA: Sinauer Associates; 2000. The Eukaryotic Cell Cycle. Available from: https://www.ncbi.nlm.nih.gov/books/NBK9876/
12. Corneo G, Ginelli E, Polli E. A satellite DNA isolated from human tissues. *Journal of Molecular Biology*. 1967 Feb 14;23(3):619–22.
13. Corneo G, Ginelli E, Polli E. Isolation of the complementary strands of a human satellite DNA. *Journal of Molecular Biology*. 1968 Apr 14;33(1):331–5.
14. Cramer P. Organization and regulation of gene transcription. *Nature*. 2019 Sep;573(7772):45–54.
15. Crick F. Central dogma of molecular biology. *Nature*. 1970; 227:561–563.
16. Eckert KA, Hile SE. Every microsatellite is different: Intrinsic DNA features dictate mutagenesis of common microsatellites present in the human genome. *Molecular Carcinogenesis*. 2009 Apr;48(4):379–88.
17. ENCODE Project Consortium. An integrated encyclopedia of DNA elements in the human genome. *Nature*. 2012 Sep 6;489(7414):57–74.
18. Hatje K, Mühlhausen S, Simm D, Kollmar M. The protein-coding human genome: Annotating high-hanging fruits. *Bioessays*. 2019 Nov;41(11):e1900066.
19. Jaekel A, Lill P, Whitelam S, Saccà B. Insights into the structure and energy of DNA nanoassemblies. *Molecules*. 2020 Nov 24;25(23):5466.

20. Junyent S, Meglicki M, Vetter R, Mandelbaum R, et al. The first two blastomeres contribute unequally to the human embryo. *Cell.* 2024 May 23;187(11):2838–2854.e17.
21. Kit S. Equilibrium sedimentation in density gradients of DNA preparations from animal tissues. *Journal of Molecular Biology.* 1961 Dec;3:711–6.
22. Kitts A, Sherry S. The Single Nucleotide Polymorphism Database (dbSNP) of nucleotide sequence variation. 2002 Oct 9 [Updated 2011 Feb 2]. In: McEntyre J, Ostell J, eds. *The NCBI Handbook* [Internet]. Bethesda, MD: National Center for Biotechnology Information. Chapter 5. Available from: https://www.ncbi.nlm.nih.gov/books/NBK21088/
23. Liehr T. Repetitive elements in humans. *International Journal of Molecular Sciences.* 2021; 22(4):2072.
24. Manders F, van Boxtel R, Middelkamp S. The dynamics of somatic mutagenesis during life in humans. *Frontiers in Aging.* 2021 Dec 17;2:802407.
25. Meselson M, Stahl FW, Vinograd J. Equilibrium sedimentation of macromolecules in density gradients. *Proceedings of the National Academy of Sciences of the United States of America.* 1957 Jul 15;43(7):581–8.
26. Nurk S, Koren S, Rhie A, Rautiainen M, Bzikadze AV, et al. The complete sequence of a human genome. *Science.* 2022 Apr;376(6588):44–53.
27. Panijpan B. The buoyant density of DNA and the G + C content. *Journal of Chemical Education.* 1977 Mar;54(3):172–3.
28. Pech M, Igo-Kemenes T, Zachau HG. Nucleotide sequence of a highly repetitive component of rat DNA. *Nucleic Acids Research.* 1979 Sep 25;7(2):417–32.
29. Piovesan A, Pelleri MC, Antonaros F, Strippoli P, Caracausi M, Vitale L. On the length, weight and GC content of the human genome. *BMC Research Notes.* 2019 Feb 27;12(1):106.
30. Preston BD, Albertson TM, Herr AJ. DNA replication fidelity and cancer. *Seminars in Cancer Biology.* 2010 Oct;20(5):281–93.
31. Rampersad S, Tennant P. Replication and expression strategies of viruses. *Viruses.* 2018:55–82.
32. Rolfe R, Meselson M. The relative homogeneity of microbial DNA. *Proceedings of the National Academy of Sciences of the United States of America.* 1959 Jul;45(7):1039–43.
33. Singer MF. Highly repeated sequences in mammalian genomes. *International Review of Cytology.* 1982;76:67–112.
34. Skinner DM. Satellite DNA's in the crabs *Gecarcinus lateralis* and *Cancer pagurus. Proceedings of the National Academy of Sciences of the United States of America.* 1967 Jul;58(1):103–10.
35. Thakur J, Packiaraj J, Henikoff S. Sequence, chromatin and evolution of satellite DNA. *International Journal of Molecular Science.* 2021 Apr 21;22(9):4309.
36. Traut TW. Physiological concentrations of purines and pyrimidines. *Molecular and Cellular Biochemistry.* 1994 Nov 9;140(1):1–22.
37. Wu Q, Bazzini AA. Translation and mRNA stability control. *Annual Review of Biochemistry.* 2023 Jun 20;92:227–45.

4

RNAs Are the Sense Molecules of Life

4.1 The Central Dogma of Molecular Biology

4.1.1 CRICK'S VISION OR "THE PSYCHOLOGICAL DRIVE"

At the "Central Dogma Lecture" on September 19, 1957, given at a Society for Experimental Biology symposium on the Biological Replication of Macromolecules at University College London, Francis Crick (1916–2004) opened with the following remark: "I shall . . . argue that the main function of the genetic material is to control (not necessarily directly) the synthesis of proteins. There is a little direct evidence to support this, but to my mind, the psychological drive behind this hypothesis is at the moment independent of such evidence."

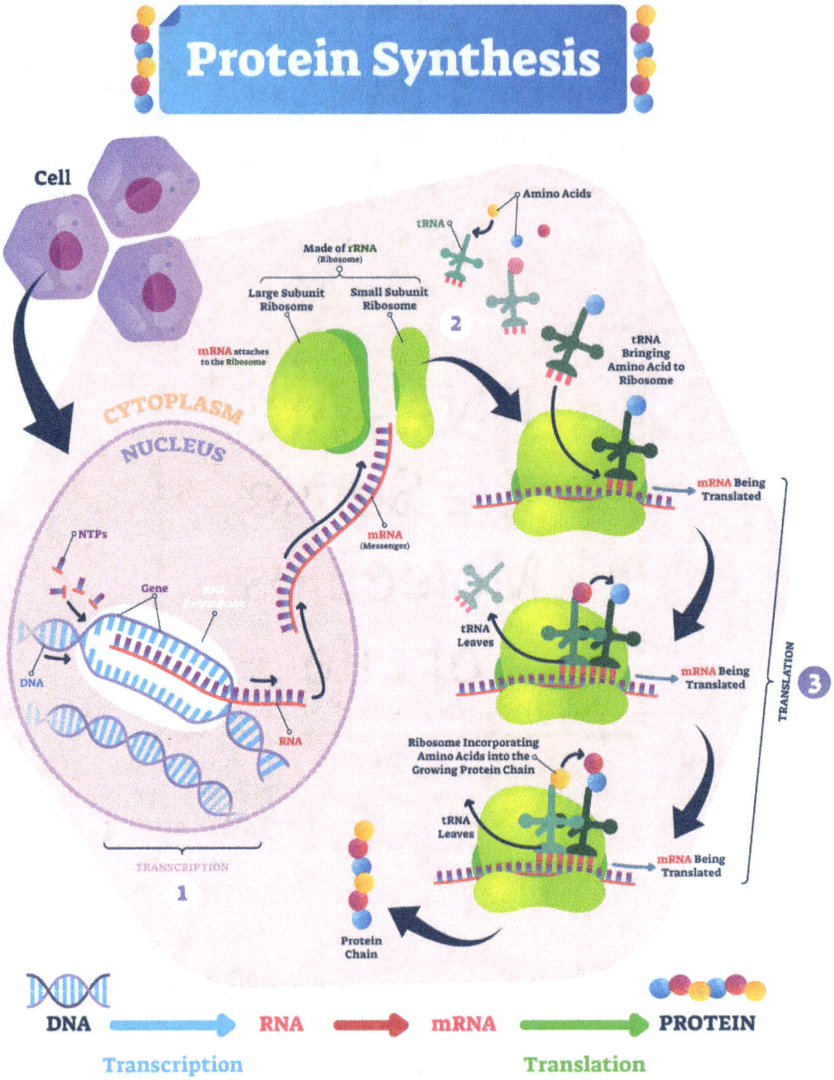

VectorMine/Shutterstock.com

FIGURE 4-1. The central dogma of molecular biology

The central dogma concerns the flow of sequence text information from one type of macromolecule to another (**Figure 4-1**). Today, it includes DNA replication (DNA to DNA), gene transcription (DNA to RNA), reverse transcription (RNA to DNA), RNA-dependent RNA synthesis (RNA to RNA, also referred to as transcription), and, centrally most important, translation (from RNA to protein/polypeptide). Crick's focus on the central dogma was primarily the terminal expresser of the information of life, the proteins. In his 1970 *Nature* article mapping the possible directions of the information transfer, he captioned it with: "The central dogma of molecular biology deals with the detailed residue-by-residue transfer of sequential information. It states that such information cannot be transferred from protein to either protein or nucleic acid."

4.1.2 STRUCTURAL DIFFERENCES BETWEEN NUCLEIC ACIDS AND PROTEINS

Crick succinctly summarized the differences between the nucleic acids and the proteins in terms of the way sequence text information is kept and transferred. The double-stranded DNA is the most common form of the genetic information carrier, and its information-carrying part, the nucleobases, can be exposed when the two strands are separated. The strand separation, most importantly, is perfectly reversible, either in vivo in the cell or in vitro in the test tube. The forces that hold the strands together in the double helix, that is, the hydrogen bonds between the base pairs, are weak forces. The weak forces allow the easy separation, but a restoration of the original form would require the forces to be cooperative. Long stretches of base pairing provide the cooperativity. In the in vivo setting, these separation and restoration events occur during gene transcription. In the in vitro setting, the separation and reassociation events are the foundations of all molecular biotechnological methods. As we have discussed earlier, base pairing is the mechanism, with the aid of the DNA or RNA polymerases to improve the efficiency, by which the sequence text information is transferred with high fidelity from a nucleic acid to another nucleic acid.

The three-dimensional structures of the biologically synthesized and active proteins are also held together by weak forces, chiefly hydrogen bonds in the secondary structures known as α-helices and β-sheets, and hydrophobic interactions that force the polypeptide to fold in the typical globular shape. Whereas breaking these forces to change the shape of the proteins to perform a different function is common, with chemical modifications of the side chains or changes of the protein solution pH or ionic strength, doing so in large portions would result in the precipitation of the protein out of solution. Unfolded polypeptides from native proteins refold over the full length, while the native proteins fold parts at a time along the biosynthesis. In some cases, the nascent polypeptides are protected by specialized proteins called chaperones to prevent the final folding until the proper working conditions are set. The best example of this is the synthesis, folding, and functions of a steroid hormone receptor, such as the androgen receptor (AR). The full-length AR is made but prevented from the final folding by the binding of heat-shock proteins (HSP70 and HSP90). The binding of dihydrotestosterone (DHT) sheds off the heat-shock proteins, and the DHT-AR complex translocates to the nucleus to activate hormonally regulated genes.

A full-length unfolding, known as denaturation, of a protein would create many possible ways to refold, and the lowest energy state achieved in the refolding is not the native state, which is achieved via partial and gradual folding. There are exceptions to this, such as the pancreatic RNase A and the Taq DNA polymerase, for which the refolded states coincide with the native states. For most proteins, unfolding to reveal the amino acid residues for possibly "making a copy" would lead to the death of the protein. Secondarily, there is no mechanism such as the base pairing for nucleic acids to allow for a totally passively dictated chemistry or biochemistry to make such copies. These are the two reasons behind Crick's remark on his vision of the central dogma.

4.2 The RNA Repertoire—Messenger RNA and Transfer RNA

At the time when Crick envisioned and conceptualized the central dogma, the transfer of genetic information was generalized as between nucleic acids and protein. The term *nucleic acids* would include both DNA and RNA. The first person to discover DNA was Friedrich Miescher in 1869 with his isolation of "nuclein" from the white blood cells in the pus on wound bandages, in the work that he published in 1871. The experimental protocol Miescher used to isolate nuclein did not produce the ultrapure DNA that the later modifications of his protocol would. But Miescher quickly realized that his nuclein was associated with other substances described by him as "contaminants," proteins that he called "protamine." The German pathologist and histologist Richard Altmann (1852–1900) used proteases to help remove the proteins and obtained rather pure DNA and gave it the name nucleic acids. Without the most intimate chemical structural understanding on nuclein and protamine, Miescher speculated that these macromolecules are capable of storing hereditary variation via stereochemistry and transmitting such information encoded in the different classes of molecules.

4.2.1 HOW IS THE RESIDUE-BY-RESIDUE SEQUENCE INFORMATION TRANSFERRED FROM DNA TO PROTEIN?

The first mechanistic model of the sequence text information transfer between the nucleic acids and the proteins was a direct complementation between the two classes of macromolecules, proposed in 1954 by the Soviet-American physicist George Gamow (1904–1968). Gamow was inspired by the double-helical structure of DNA determined and published in 1953 by Watson and Crick and contemplated on how a text written with four letters, ACGT in the DNA, can encode a text written with 20 letters, the amino acids that make up a polypeptide in a protein. He drew up a model in which the four bases in the two adjacent base pairs would form holes of different sizes and shapes to accommodate the 20 different amino acids of different sizes and shapes into a line for a co-linear manner of biochemical synthesis. The combinations of changing three of the four bases surrounding the hole would give the requisite varieties for 20 different amino acids. This model as we now know is incorrect, but nonetheless quite ingenious, no, brilliant for the time.

Francis Crick in his "ON DEGENERATE TEMPLATES AND THE ADAPTOR HYPOTHESIS—A Note for the RNA Tie Club," hand-marked "early 1955," quoted Kai Kaus ibn Iskandar: "Is there anyone so utterly lost as he that seeks a way where there is no way." The RNA Tie Club was a 20-member informal science club of molecular biologists, representing the 20 amino acids, suggested by George Gamow in 1953 to "solve the riddle of the RNA structure and to understand how it built proteins" and to "do or die; or don't try." The members wore a tie with embroidered nucleic acid helices, true to the name of the club. This super brief introduction of the historical background should prepare the readers today for this remark by Crick in the Note: "It is convenient to start with some criticisms of Gamow's paper as they lead naturally to the further points I wish to make." With data on the insulin chains, Crick invalidated the Gamow model of DNA coding for amino acids in a polypeptide. But he also underscored the value of Gamow's line of thinking, that different base sequences can code for one amino acid, considered a new idea called "degeneracy," and that "coding" itself is the inspiration of Gamow to him and others. Crick remarked: "Finally it is obvious to all of us that without our President the whole problem would have been neglected and few of us would have tried to do anything about it."

4.2.2 AN ADAPTOR IS REQUIRED FOR THE TRANSFER OF INFORMATION FROM DNA TO PROTEIN

Crick explained how nucleic acids cannot serve as a direct or immediate template for amino acid assembly in a polypeptide, primarily because of the mismatch in structural dimensions: "If one considers the physical-chemical nature of the amino acid side chains we do not find complimentary (*sic*) features on the nucleic acid." The problem, however, can be solved if each amino acid is linked to a small molecule "adaptor," which could fit complementarily with the nucleic acid template. The "adaptor hypothesis" was coined by Sydney Brenner (1927–2019), with whom Crick discussed his idea.

The German biochemist Albrecht Kossel (1853–1927) was awarded The Nobel Prize in Physiology or Medicine in 1910 for his discovery of the nucleobases, adenine, cytosine, guanine, thymine, and uracil as components of nucleic acids. The Russian-born American biochemist Phoebus Levene (1869–1940) studied the structures of nucleic acids and determined the structural components that made up DNA and RNA, chiefly the deoxyribose and ribose with their sugar-phosphate bonds to form the backbone of the chains. Each unit in the chain is called a "nucleotide" by Levene, who proposed the tetranucleotide hypothesis of DNA containing equal portions of A, C, G, and T, disproven later by Erwin Chargaff. Levene suggested the new classification of nucleic acids by the differences of the backbone sugar ring: "It is therefore logical to classify nucleic acids according to their component carbohydrate into ribonucleic acids and ribodesose (*i.e.* deoxyribose) nucleic acids."

An independent and also earlier hypothesis on the role of RNA in the relay of the encoded genetic information from DNA to protein was proposed in 1947 by André Boivin (1895–1949): "The macro-molecular desoxyribonucleic acids govern the building of macro-molecular ribonucleic acids, and, in turn, these control the production of cytoplasmic enzymes." The initial thought of an RNA intermediate for guiding protein synthesis, specifically, amino acid residue assembly, led to the "one gene—one ribosome—one protein" hypothesis because the ribosomal RNA (rRNA) is the most abundant in a cell. The rRNAs, however, are very homogeneous across the species (in bacteria) and stable. This would not be compatible with the requirement of cells turning over old and making new proteins in response to the environmental changes. The true intermediate needs to be turned over, thus, unstable.

4.2.3 THERE IS A MESSENGER (RNA) FOR THE TRANSFER OF INFORMATION FROM DNA TO PROTEIN

In May 1961, two back-to-back papers in the same issue of *Nature*, Brenner *et al.* (including Jacob and Meselson) and Gros *et al.* (including Watson) described the messenger RNA (mRNA) as "an unstable intermediate carrying information from genes to ribosomes for protein synthesis" (title of the first paper). The leading researchers agreed to coordinate their publications to appear together for such a landmark finding in molecular biology. The model system used was T2 bacterial phage infection of *Escherichia. coli* cells, an event that shuts down the host protein synthesis and switches to the phage protein synthesis. The title of the second paper described this very accurately: "Unstable Ribonucleic Acid Revealed by Pulse Labeling of *Escherichia coli*." "Pulse labeling" is a technique of using radioactive components of biomolecules to specifically "label" the newly synthesized portions after a specific time, herein, phage infection.

4.2.4 THE ADAPTOR IS THE TRANSFER RNA

Just as DNA does not offer a ready interface for complementing the different side groups of the 20 amino acids in proteins, the mRNA also does not have a direct mechanism of doing this. The small "adaptor" molecule is the transfer RNA (tRNA), discovered in 1957 by a Harvard-MGH group led by Paul Zamecnik (1912–2009), as "A Soluble Ribonucleic Acid Intermediate in Protein Synthesis." Their work showed that by an adenosine triphosphate (ATP)–dependent enzymatic reaction, a covalent linkage between this small and soluble RNA and an amino acid is formed. The amino acid in this "intermediate carrier" can be "transferred" to the peptide linkage in a protein. In essence, the mRNA would contain the "transcribed" genetic code from the DNA and can base-complement with an adaptor RNA carrying amino acids for assembly into proteins.

Watson and his coauthors gave the first outline of the reduced-to-practice central dogma in their 1961 paper on the discovery of mRNA: "Our working hypothesis is that no fundamental difference exists between protein synthesis in phage infected and uninfected bacteria. In both cases typical ribosomal RNA does not carry genetic information, but has another function, perhaps to provide a stable surface on which transfer RNA's can bring their specific amino acids to the messenger RNA template." Over the great many years with the great many new discoveries in molecular biology, this description remained very accurate. The assumption that all forms of life, including half forms of life such as bacterial phages and viruses that require a host cell to manifest life, would obey the same biochemical principles in the genetic information transfer has proven to be true. A major and the only major difference between the prokaryotes such as *E. coli* and the eukaryotes is the compartmentalization of transcription in the nucleus and translation in the cytoplasm for the latter. As far back as the 1940s, Jean Brachet (1909–1988) in Belgium and Torbjörn Caspersson (1910–1997) in Sweden reported that RNA is found primarily in the cytoplasm, where protein synthesis takes place.

4.3 The Genetic Code

It is directly intuitive for one to look at the linear residue-by-residue sequence in nucleic acids and proteins to conjure up the parallel between the potential text information in these macromolecules of life and the text information of the written language with letters and words. The question is: What are the rules and how are the molecular letters A, C, G, and T in DNA, or A, C, G, and U in RNA put into words that match the amino acids?

4.3.1 SIMPLE MATH ON THE NUMBER OF CODONS

The first attempt at decoding a nucleic acid to account for the 20 different amino acids was that of Gamow's, with overlapping letters in the double-stranded DNA. The mRNA is a single-stranded nucleic acid with four different letters. To code for 20 different amino acids, the minimum size of the coding unit (codon) needs to be three letters: With four choices each, there can be 64 varieties; whereas a two-letter code can give only 16 kinds. This baseline calculation is not negotiable, but questions remain as to how the three-letter codons are formed along the linear sequence in an mRNA. First, there is redundancy of the 64 codons for 20 amino acids. Thus, either not all of the codons are used, or at least for some amino acids, there are more than one codon. Silent codons would have to be excluded in the DNA and the transcribed RNA, or otherwise they would render the code discontinuous. By the same line of thinking, the three-letter codons should be connected with no gaps. Second, are the codons overlapping? If they are, then the "adaptor" molecules can get on only one at a time. This means something else needs to help hold a nascent polypeptide chain in place for the transfer of the amino acid linked to the adaptor.

4.3.2 THE GENETIC CODE WAS DEDUCED FROM A CONTROL EXPERIMENT

The genetic code was decoded in the early 1960s. The approach taken was to use the protein translation machinery, chiefly the ribosomes, in the *E. coli* cell after stripping away the nucleic acids. This approach was known as in vitro translation (IVT), which has regained its popularity and glory in recent years. The first genetic codon was determined quite as an accident. Marshall Nirenberg (1927–2010) at the National Institutes of Health (NIH, Bethesda, MD) was able to incorporate radioactive amino acids into proteins by cell-free IVT. Nirenberg communicated his findings with scientists working to direct the synthesis of a plant virus coat protein by cell-free IVT. The challenge at the time was to validate the new protein being synthesized after the addition of the plant viral RNA to be the proper coat protein. Nirenberg used a synthetic RNA (sRNA) with a "nonsensical" sequence, nothing but uridine, a poly-U, as a negative control in an experiment with the intent to show that a string of U's should not code for anything, thus will not direct the synthesis of a protein.

The unintended result from this negative control experiment in 1961 was the beginning of cracking the genetic code: Not a protein but a poly-phenylalanine peptide was produced when the synthetic poly-U RNA was added to the cell-free IVT system. This result suggested that the codon for phenylalanine was uridine, but how many uridines remained a question. In the follow-up study in 1962, Nirenberg's lab went on to show that the RNA template (mRNA) needs to be single stranded. When poly-A was added, it quickly formed a double helix with the poly-U and inhibited the synthesis of poly-phenylalanine. They also demonstrated that the amino acyl phenylalanine-tRNA (sRNA at the time) is a required intermediate. Nirenberg proposed a base-complementation mechanism with adenines in the tRNA (sRNA) to pair with the poly-U template. The triplet codons were then decoded with sRNA templates of increasing variations, such as a randomly ordered poly-UC, giving the codons for phenylalanine, serine, leucine, and proline.

H Gobind Khorana (1922–2011), who shared the 1968 Nobel Prize in Physiology or Medicine with Nirenberg and Robert Holley (1922–1993), expanded the decoding work by taking advantage of the contemporary advancements in the enzymology of nucleic acid biosynthesis, the discovery and purification of *E. coli* DNA polymerase I (Arthur Kornberg) and a DNA-dependent RNA polymerase (Jerard Hurwitz). These advances and the availability of the tool enzymes overcame the challenges involved in chemically synthesizing RNA when chemically synthesizing DNA was more efficient. DNA templates with defined sequences were made and then transcribed by the RNA polymerase to produce the mRNA used for decoding the genetic code. The scheme is known as in vitro transcription and translation (IVTT).

Robert Holley's work that earned him the shared Nobel Prize was the sequencing of tRNA, using an ingenious method of digesting RNA differentially at specific nucleotides by different nucleases, such as bovine pancreatic RNase, Taka-Diastase RNase T1, and snake venom phosphodiesterase. Khorana's way of making long DNA from short synthetic DNA with overlaps and gaps to be filled with a DNA polymerase, and Holley's way of jigsaw puzzle sequencing nucleic acids via nucleotide-specific cleavages were inspirational to future advances in molecular biology methods. In his Nobel lecture in 1968, Khorana gave the now well-known codon table with the 64 combinations of the triplet codons (including the termination codons) and provided experimental evidence on how the "anticodon" in the tRNA would pair up with the codon in an mRNA template. The genetic code is triplet, nonoverlapping, with no gap in the template; and nearly universal for all forms of life.

4.4 Ribosomal RNA

The 1974 Nobel Prize in Physiology or Medicine was shared by Albert Claude (1899–1983), Christian de Duve (1917–2013), and George Palade (1912–2008) "for their discoveries concerning the structural and functional organization of the cell." Palade discovered the ribosomes, formerly known as "Palade granules," a description of the granular appearance of these structures. In 1958, Howard Dintzis (1927–2024) renamed these as "ribosomes," for "ribonucleoprotein particles of the microsome fraction" when the cellular contents were fractionated by biophysical means. Palade pioneered the study on the endoplasmic reticulum and the ribosomes for their role in the eukaryotic cellular protein synthesis and showed the itinerary and pathways of production and packaging of the secretory proteins through the intracellular membrane networks.

Ribosomes are "ribonucleoprotein particles" by the original descriptions upon their discovery, and the RNA components in the ribosomes make up approximately 60% of these particles. In the total cellular RNA, approximately 80% are rRNA, which are identified by their sedimentation coefficients. The prokaryotic cells contain the 5S, 16S, and 23S rRNA. The eukaryotic cytoplasmic rRNAs are 5S, 17S or 18S, and 26S or 28S; and a special 5.8S rRNA, corresponding to the 5'-terminal 160 nucleotides of the prokaryotic 23S rRNA.

The ribosome is the machinery for protein synthesis in all forms of cellular life, and more than 60% of the cell's energy consumption is required for the biogenesis of ribosomes. The components and structures of the ribosomes are remarkably conserved from bacteria to humans, especially the functional sites responsible for protein synthesis: the decoding center (DC) and the peptidyl transferase center (PTC). The essential role of the ribosome in protein translation requires precision in the assembly of its subunits, which are highly complex. The human 80S ribosome is 4,300 kDa in molecular weight, composed of 80 ribosomal proteins and four rRNAs. Problems in the processing and folding of the rRNAs and the assembly of the ribosomal subunits are associated with genetic disorders, collectively known as ribosomopathies. Overall, there are not very many such ribosomopathies because major problems in ribosomal assembly would be catastrophic to the cell.

4.5 All Other RNAs

The RNA world is once again expanding, as it is believed to be the first molecule of life on earth owing to its compatibility with the base-pairing mechanism of replication, its ability to assume a three-dimensional structure, and its 2'-OH (hydroxyl) group that can act as an enzyme, ribozyme. The 1989 Nobel Prize in Chemistry was shared by Sidney Altman (1939–2022) and Thomas Cech for "their discovery of catalytic properties of RNA." In the late 1980s to the late 1990s, the doctrine of tightly controlled gene expression was dominating and the prevailing view was that any and all RNA transcripts were produced in the cell for a reason or purpose. The rRNA and tRNA were thought to be nearly perfectly understood for their biological roles and functions. Despite their proportions in the total cellular RNA content, at roughly 80% (rRNA) and 15% (tRNA), respectively, the focus of research attention was keenly on the mRNA, which is less than 4% of the total cellular RNA. A typical photographed image of agarose gel electrophoresis analysis of the total cellular RNA would show two distinctive fluorescent dye-stained major bands of rRNA (28S and 18S) from the large and small subunits of the ribosome. There is a background smear, which is the mRNA of varying sizes. At the bottom of the gel in each sample lane, there is a blob of nucleic acids, where the relatively small tRNA (76–90 nucleotides) would migrate. Above the tRNA blob, the 5.8S rRNA may also aggregate as a band.

The investigation of mRNA species and abundance, and especially their changes in response to the changes in the environment in which the cell resides, is the most central to the theme of the central dogma. Over the years, we described this as cell-specific, tissue-specific, and condition-specific gene expression. Today, it is called "transcriptomics," as the molecular representation in the genetic information transfer, in quality and quantity, also a fun wordplay of the originally meaningful word "genomics." rRNA and tRNA are also part of this central dogma, but they are "noncoding" and "terminally functional" RNA without an encoded protein product downstream.

The attention on noncoding RNAs increased with the 2006 Nobel Prize in Physiology or Medicine to Andrew Fire and Craig Mello "for their discovery of RNA interference—gene silencing by double-stranded RNA." The class of RNA that they discovered is called microRNA (miRNA), capable of binding to mRNA to block protein translation. The latest buzz is long noncoding RNA (lncRNA), which are believed to be previously "junk transcripts" that gradually gained a function in regulating cellular biology. The endowment of such gained functions also revises the long-held view of very tightly regulated transcription at sites of opened double-stranded DNA. RNA polymerases will initiate transcription at any open and accessible DNA, and the current estimate is that 80% of the human genome is transcribed, that is, an RNA transcript has been identified with a matching genomic DNA sequence, but only less than 2% code for proteins.

4.6 Gene and RNA Transcript Copy Numbers

The typical figure illustration in a genetics or molecular biology textbook would show a double-stranded DNA molecule as the gene, with one mRNA molecule transcribed off the DNA gene and with one polypeptide (protein) molecule translated from the mRNA. But the quantitative relationships along and across the routes of the central dogma are very important to our understanding of molecular biology and genetics.

4.6.1 GENE COPY NUMBERS

At the DNA/gene level for a diploid cell, such as a human cell, there are two <u>copies</u> of each single-<u>copied</u> gene. We use the term "single-copied gene" to describe those that do not have a duplicated copy. The word "copy" has very different meanings in the two uses underscored above, but is unfortunately used for both instances. In the first use, it is equivalent to "allele," and in the second, it is true to the sense, that is, a copy of a gene at a new locus. For most genes, both copies/alleles are used for transcription, known as biallelic expression. For some "imprinted genes," one copy/allele from the imprinting parent is not used for transcription. Transcription is limited to the un-imprinted copy/allele from the other parent, known as monoallelic expression.

4.6.2 THE MOST ABUNDANT RNA SPECIES IN THE CELL—rRNA AND tRNA

For the rRNA of humans, the genes are duplicated and located as clusters on the short arms of acrocentric chromosomes 13, 14, 15, 21, and 22. The human rRNA genes are transcribed by RNA polymerase I into the 47S rRNA precursor, which is processed to yield the 18S, 5.8S, and 28S rR-

NAs. The 5S rRNA gene copies in humans were previously estimated to be 1,700 to 2,000, but newer estimates lowered the number to 18. There are several hundred copies/duplicates (loci) of the rRNA genes being transcribed in each actively dividing cell, producing several million transcripts *per* cell cycle to maintain a steady-state level of several million ribosomes *per* cell.

For the tRNA of humans, there are 619 predicted genes of 57 anticodon families, with unequal copy numbers/duplicates (loci) and 432 unique tRNA transcripts. The tRNA genes are transcribed by RNA polymerase III. The tRNA is the most abundant type of RNA in the cells by transcript numbers, with tens of millions, and the highest by molar concentration, at up to 100 μM.

4.6.3 THE ABUNDANCE OF CELLULAR MESSENGER RNA TRANSCRIPTS VARIES TREMENDOUSLY

It makes the perfect sense for the rRNA and tRNA to be present at highly abundant and constant levels because of their roles as the assembly platform and workers of protein synthesis. It would not make much sense for the mRNA of each protein-coding gene to be produced at equally high and constant levels. The mRNA was discovered as an unstable RNA species. This makes more sense as the cell's needs in the types of proteins and their quantities change when the cell's environment changes, which would demand different proteins performing different roles.

The biggest human gene, DMD, encoding the structural protein dystrophin expressed in muscle cells, at ~2.4 Mbp (million base pairs), requires 16 hours to transcribe, by RNA polymerase II. The dystrophin mRNA is very stable, with a half-life of about 16 hours, and only about 5 to 10 transcripts *per* cell. This is an example of a low-copy number transcript. The human β-globin mRNA is also rather stable, with a half-life of 10 to 20 hours. Data from studies in rabbit proerythroblasts indicate that the β-globin mRNA is present at a level of 10,000 to 23,000 transcripts *per* cell during the cell cycle. This high level of abundance enabled the shortcut to clone and sequence the β-globin mRNA (cDNA) from rabbit reticulocytes.

On the other hand, for material supply during the G1 and S phases of the cell cycle in preparation for DNA replication, the histone protein abundance needs to increase by hundreds of fold to ensure that the nascent DNA is immediately bound to histones. This is helped by the ~35-fold increase in transcription. Indeed, the histone mRNA is the most abundant during the S phase. The absolute material needed for the DNA replication *per* cell cycle is only one-fold extra, and the highly redundant histones need to be destroyed at the end of DNA replication, as well as the highly abundant histone mRNA. It then makes the perfect sense that the histone mRNA is the only kind that does not have a poly-A tail, with a half-life of less than 1 hour during the S phase and reduced to 10 minutes at the end of the S phase.

BIBLIOGRAPHY

1. https://www.nobelprize.org/prizes/medicine/1910/kossel/facts/
2. Amaral P, Carbonell-Sala S, De La Vega FM, et al. The status of the human gene catalogue. *Nature*. 2023; 622:41–7.
3. Anderson WF. The effect of tRNA concentration on the rate of protein synthesis. *Proceedings of the National Academy of Sciences of the United States of America*. 1969 Feb;62(2):566–73.
4. Azad AA, Zoubeidi A, Gleave ME, Chi KN. Targeting heat shock proteins in metastatic castration-resistant prostate cancer. *Nature Reviews Urology*. 2015 Jan;12(1):26–36.
5. Beaudet AL. Differences in RNA codon recognition as a function of cellular tRNA content, 1967. Yale Medicine Thesis Digital Library, p. 2384.
6. Bernfield MR, Nirenberg MW. RNA codewords and protein synthesis. The nucleotide sequences of multiple codewords for phenylalanine, serine, leucine, and proline. *Science*. 1965 Jan 29;147(3657):479–84.
7. Brenner S, Jacob F, Meselson M. An unstable intermediate carrying information from genes to ribosomes for protein synthesis. *Nature*. 1961 May 13;190:576–81.
8. Chan PP, Lin BY, Mak AJ, Lowe TM. tRNAscan-SE 2.0: Improved detection and functional classification of transfer RNA genes. *Nucleic Acids Research*. 2021 49:9077–96.
9. Clissold PM, Arnstein HR, Chesterton CJ. Quantitation of globin mRNA levels during erythroid development in the rabbit and discovery of a new beta-related species in immature erythroblasts. *Cell*. 1977 Jun;11(2):353–61.
10. Cobb M. Who discovered messenger RNA? *Current Biology*. 2015 Jun 29;25(13):R526–32.
11. Cooper GM. *The cell: A molecular approach*. 2nd ed. Sunderland, MA: Sinauer Associates; 2000. The Nucleolus. Available from: https://www.ncbi.nlm.nih.gov/books/NBK9939/
12. Crick FH. On degenerate templates and the adaptor hypothesis: A note for the RNA Tie Club. 1955.
13. Crick FH. Codon–anticodon pairing: The wobble hypothesis. *Journal of Molecular Biology*. 1966 Aug;19(2):548–55.
14. Dintzis HM. Assembly of the peptide chains of hemoglobin. *Proceedings of the National Academy of Sciences of the United States of America*. 1961 Mar 15;47(3):247–61.
15. Frixione E, Ruiz-Zamarripa L. The "scientific catastrophe" in nucleic acids research that boosted molecular biology. *Journal of Biological Chemistry*. 2019 Feb 15;294(7):2249–55.
16. Fry M. Crick's adaptor hypothesis and the discovery of transfer RNA: Experiment surpassing theoretical prediction. *Philosophy, Theory, and Practice in Biology*. 2022;14(11):1–31.
17. Gamow G. Possible relation between deoxyribonucleic acid and protein structures. *Nature*. 1954; 173:318.

18. Giuliano MG, Engl C. The lifecycle of ribosomal RNA in bacteria. In: Kotta-Loizou I, ed. *RNA damage and repair*. Cham, CH: Springer; 2021 Jul. Chapter 2. Available from: https://www.ncbi.nlm.nih.gov/books/NBK594409/
19. Gros F, Hiatt H, Gilbert W, Kurland CG, Risebrough RW, Watson JD. Unstable ribonucleic acid revealed by pulse labelling of *Escherichia coli*. *Nature*. 1961 May 13;190:581–5.
20. Haselkorn R. Poly(U) cracked the code: A Marshall Nirenberg memoir. *FASEB Journal*. 2010 Apr;24(4):968–9.
21. Hoagland MB, Stephenson ML, Scott JF, Hecht LI, Zamecnik PC. A soluble ribonucleic acid intermediate in protein synthesis. *Journal of Biological Chemistry*. 1958 Mar;231(1):241–57.
22. Holley RW, Madison JT, Zamir A. A new method for sequence determination of large oligonucleotides. *Biochemical and Biophysical Research Communications*. 1964;17(4):389–94.
23. Holley RW, Apgar J, Everett GA, Madison JT, Marquisee M, Merrill SH, Penswick JR, Zamir A. Structure of a ribonucleic acid. *Science*. 1965 Mar 19;147(3664):1462–5.
24. Holley RW, Everett GA, Madison JT, Zamir A. Nucleotide sequences in the yeast alanine transfer ribonucleic acid. *Journal of Biological Chemistry*. 1965 May;240:2122–8.
25. Hori Y, Engel C, Kobayashi T. Regulation of ribosomal RNA gene copy number, transcription and nucleolus organization in eukaryotes. *Nature Reviews Molecular Cell Biology*. 2023; 24, 414–29.
26. Iben JR, Maraia RJ. tRNA gene copy number variation in humans. *Gene*. 2014 Feb 25;536(2):376–84.
27. Johnston R, Aldrich A, Lyons SM. Roles of ribosomal RNA in health and disease. *Frontiers in RNA Research*. 2024;1:1331185.
28. Khorana HG. Synthesis in the study of nucleic acids. The Fourth Jubilee Lecture. *Biochemistry Journal*. 1968 Oct;109(5):709–25.
29. Lamm E, Harman O, Veigl SJ. Before Watson and Crick in 1953 Came Friedrich Miescher in 1869. *Genetics*. 2020 Jun;215(2):291–6.
30. Levene PA. The structure of yeast nucleic acid IV. ammonia hydrolysis. *Journal of Biological Chemistry*. 1919;40:415–24.
31. Marzluff WF, Duronio RJ. Histone mRNA expression: Multiple levels of cell cycle regulation and important developmental consequences. *Current Opinion in Cell Biology*. 2002 Dec;14(6):692–9.
32. Nakhoul H, Ke J, Zhou X, Liao W, Zeng SX, Lu H. Ribosomopathies: mechanisms of disease. *Clinical Medicine Insights: Blood Disorders*. 2014 Aug 14;7:7–16.
33. Nirenberg MW, Matthaei JH. The dependence of cell-free protein synthesis in *E. coli* upon naturally occurring or synthetic polyribonucleotides. *Proceedings of the National Academy of Sciences of the United States of America*. 1961 Oct 15;47(10):1588–602.
34. Nirenberg MW, Matthaei JH, Jones OW. An intermediate in the biosynthesis of polyphenylalanine directed by synthetic template RNA. *Proceedings of the National Academy of Sciences of the United States of America*. 1962 Jan 15;48(1):104–9.

35. Noller HF. Structure of ribosomal RNA. *Annual Review of Biochemistry*. 1984;53:119–62.
36. Palade GE. A small particulate component of the cytoplasm. *Journal of Biophysical and Biochemical Cytology*. 1955 Jan;1(1):59–68.
37. Palazzo AF, Lee ES. Non-coding RNA: What is functional and what is junk? *Frontiers in Genetics*. 2015 Jan 26;6:2.
38. Pan T. Modifications and functional genomics of human transfer RNA. *Cell Research*. 2018 Apr;28(4):395–404.
39. Panse VG, Johnson AW. Maturation of eukaryotic ribosomes: Acquisition of functionality. *Trends in Biochemical Science*. 2010 May;35(5):260–6.
40. Park EG, Ha H, Lee DH, Kim WR, Lee YJ, Bae WH, Kim HS. Genomic analyses of non-coding RNAs overlapping transposable elements and its implication to human diseases. *International Journal of Molecular Science*. 2022 Aug 11;23(16):8950.
41. Peixeiro I, Silva AL, Romão L. Control of human beta-globin mRNA stability and its impact on beta-thalassemia phenotype. *Haematologica*. 2011 Jun;96(6):905–13.
42. Strauss BS. Martynas Yčas: The "Archivist" of the RNA Tie Club. *Genetics*. 2019 Mar;211(3):789–95.
43. Tamura K. Origins and early evolution of the tRNA molecule. *Life* (Basel). 2015 Dec 3;5(4):1687–99.
44. Tennyson CN, Shi Q, Worton RG. Stability of the human dystrophin transcript in muscle. *Nucleic Acids Research*. 1996 Aug 1;24(15):3059–64.
45. Vierna J, Wehner S, Höner zu Siederdissen C, et al. Systematic analysis and evolution of 5S ribosomal DNA in metazoans. *Heredity*. 2013 111:410–421.
46. Walter NG, Engelke DR. Ribozymes: Catalytic RNAs that cut things, make things, and do odd and useful jobs. *Biologist* (London). 2002 Oct;49(5):199–203.
47. Watson J, Crick F. Genetical implications of the structure of deoxyribonucleic acid. *Nature*. 1953; 171:964–7.
48. Yusupova G, Yusupov M. Crystal structure of eukaryotic ribosome and its complexes with inhibitors. *Philosophical Transactions of the Royal Society of London. Series B, Biological Sciences*. 2017 Mar 19;372(1716):20160184.
49. Zhou X, Liao WJ, Liao JM, Liao P, Lu H. Ribosomal proteins: Functions beyond the ribosome. *Journal of Molecular Cell Biology*. 2015 Apr;7(2):92–104.

5

Genetic Variants Affect Protein Structure and Function

5.1 One Gene, One Enzyme (Protein)

The credit of establishing that DNA is the genetic material to "transform," that is, giving new biological properties to a form of life was given to Fred Griffith (1877–1941), with his 1928 publication in which he used heat-inactivated extracts of *Pneumococcus* strains to transform others. Griffith's work was far from complete in support of the final conclusion that DNA is the genetic material, and this conclusion was also "read back" into his experiments, very much like "Mendel's laws of genetics" being read back to Mendel's work. The results from some of Griffith's experiments can actually be used to make a better argument for proteins to be the transforming agent.

Further work in 1944 by Oswald Avery (1877–1955), Colin MacLeod (1909–1972), and Maclyn McCarty (1911–2005) using purified DNA in the transformation experiments showed that DNA can transform, but did not rule out proteins for being able to achieve the same. In 1952, Alfred Hershey (1908–1997) and Martha Chase (1927–2003) in their famous "blender experiment" used differential labeling of the protein coat (with ^{35}S) and the DNA core (with ^{32}P) of the T2 bacteriophage to achieve a cleaner separation of protein and DNA for the determination of the genetic material. Bacteriophages infect *Escherichia coli* cells by adsorption to the cells, Hershey and Chase used a household blender to shake the phages off the cells after the adsorption and found that the bulk (75%) of the ^{35}S-labeled material (protein) was outside the cell, along with only a small fraction (15%) of the ^{32}P-labeled material (DNA). More telling was the result that the new phages produced by the infected cells contained less than 1% of the starting ^{35}S-labeled material (protein) but greater than 30% of the starting ^{32}P-labeled material (DNA). This was the first time that a form of life (dependent on a host cell) with a simple protein-DNA makeup was shown to use its DNA part as the material of transmission in the next generation, and most likely not involving its protein part. Equally important was the observation that the passage of this form of life involved the material that contained the text. These principles were extrapolated to all forms of life based on nucleic acids and proteins, and the extrapolation has been proven true. We are confident today that proteins cannot serve as the genetic material for the passage of life, as Crick stated in his framing of the central dogma. But scientifically speaking, the only evidence we can produce was a failure for proteins to do so in any attempt.

The term "gene" to describe the unit of heredity was from the German word Gen, first used in 1909 by the Danish botanist Wilhelm Ludvig Johannsen (1857–1927). Three years before the Avery-MacLeod-McCarty experiments and 10 before the Hershey and Chase experiments, George Beadle (1903–1989) and Edward Tatum (1909–1975) were credited with the "one gene, one enzyme" hypothesis. In their 1941 paper in *Proceedings of the National Academy of Sciences* (*PNAS*, USA), Beadle and Tatum evaluated the research investigations into the biochemical mechanisms of the physiologically manifested phenotypes, for example, fermentation of sugars by yeasts. They pointed out that the approach at the time, akin to "forward genetics," was limited to the nonlethal heritable characters, likely nonessential "terminal" characters, leading to the contemporary belief that genes control only "superficial" characters. This approach would be powerless for the investigation of normal enzymatic pathways that do not lead to a manifested phenotype. In essence, there was no change or variation to help flag a biochemical reaction or pathway. A genetic disorder, such as alkaptonuria, can produce the abnormal accumulation of a metabolite and indicate the loss of an enzyme.

The Beadle and Tatum approach "to investigate the general problem of the genetic control of developmental and metabolic reactions by reversing the ordinary procedure" was to use X-ray to generate mutants in the bread mold *Neurospora crassa* or *Neurospora sitophila*, akin to "reverse genetics." A mutant with a failed biochemical reaction would not be able to produce the next product, and if that product is essential, the mutant would not be able to survive until the product is resupplied.

In the about 2,000 strains derived from the X-rayed parental strains, three mutants grew normally on the complete medium, but not on the minimal medium, with only sucrose as the carbon source. Subsequent growth tests on the minimal medium with the addition of known vitamins identified an *N. sitophila* strain as not being able to synthesize vitamin B_6 (pyridoxine) and another as not being able to synthesize vitamin B_1 (thiamine). An *N. crassa* strain was unable to synthesize para-aminobenzoic acid. The complementation of the minimal medium with a diffusing gradient of vitamin B_6 showed that the vitamin B_6 mutant can grow as fast as the normal strain, with the rate being dictated by the vitamin B_6 concentration. Crossing with the normal strain produced results to further suggest that the mutation occurred in a single gene. These X-ray mutagenesis and genetic crossing experiments were in principle not different from the fruit fly experiments carried out in the Morgan laboratory, in which Beadle worked between 1931 and 1935. The fly phenotypes generated would have fallen under "nonlethal, nonessential, and superficial," and the studies did not directly reduce the phenotype to an enzymatic pathway. Beadle and Tatum shared the 1958 Nobel Prize in Physiology or Medicine "for their discovery that genes act by regulating definite chemical events," along with Joshua Lederberg (1925–2008), "for his discoveries concerning genetic recombination and the organization of the genetic material of bacteria."

5.2 The Classes of Proteins

Today, based on the structures and functions, we can catalog the naturally occurring proteins in the following classes.

5.2.1 ENZYMES

These proteins are catalysts of biochemical reactions. They participate in the conversion of molecules to meet the growth and development requirements of biological lives. We have discussed the alcohol dehydrogenases and aldehyde dehydrogenases in the metabolism of beverage alcohol. The DNA polymerases involved in replication and repair, and the RNA polymerases involved in transcription are synthetic enzymes. The nucleases used in molecular biology methods to aid the determination of RNA or DNA sequences are degradative enzymes, as are the proteases such as trypsin, made abundantly in the pancreas for secretion into the intestines to digest proteins in food.

Some proteases cleave other proteins not in a degradative manner, but in a modification manner, called a limited proteolysis. The best known examples of these are the serine proteases involved in blood coagulation (**Figure 5-1**). The transfer of genetic information from DNA to messenger RNA (mRNA) to protein is a spatially and temporally regulated process. The production of a protein requires a certain amount of time from the moment the protein-coding gene is activated upon a proper signaling to the cell, such as the stimulation by a growth factor. For blood coagulation, time is of essence and there is none when there is a tissue and blood vessel damage. If the process starts in the liver with the transcription of the coagulation protein genes, there is only one outcome, bleeding to death. Thus, in our bloodstream, there are coagulation proteins already made to do the job, but cannot be active, either, or blood clots would form, known as thrombosis. The solution is to make these proteins in a precursor form, inactive, but readily activated upon a single-step proteolytic cleavage.

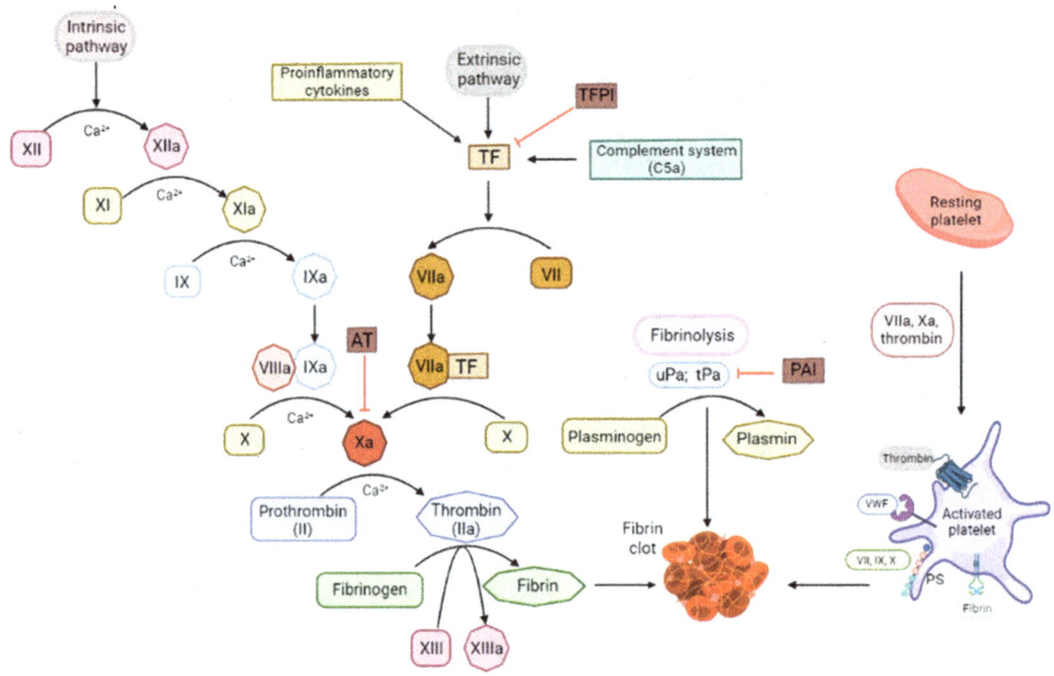

FIGURE 5-1. The blood coagulation cascade is activated by limited proteolysis involving serine proteases.

5.2.2 STRUCTURAL PROTEINS AND MOTOR PROTEINS

These proteins form the structural support of cells and work to change the cell shape in response to mechanical impact. The best example is the protein dystrophin (**Figure 5-2**), named after the hereditary Duchenne muscular dystrophy (DMD), an X-linked recessive disease affecting mainly boys of young age. With an analogy to a part of an automobile, the structure and function of dystrophin, a very large structural protein of the muscle cells, resemble that of the shock absorber, a molecular shock absorber. It connects the cytoskeleton to the extracellular matrix (basal lamina) and endows the muscles with elasticity during contraction. Myosin, dynein, and kinesin are examples of motor proteins, their motor actions are also dependent on interactions with the structural proteins.

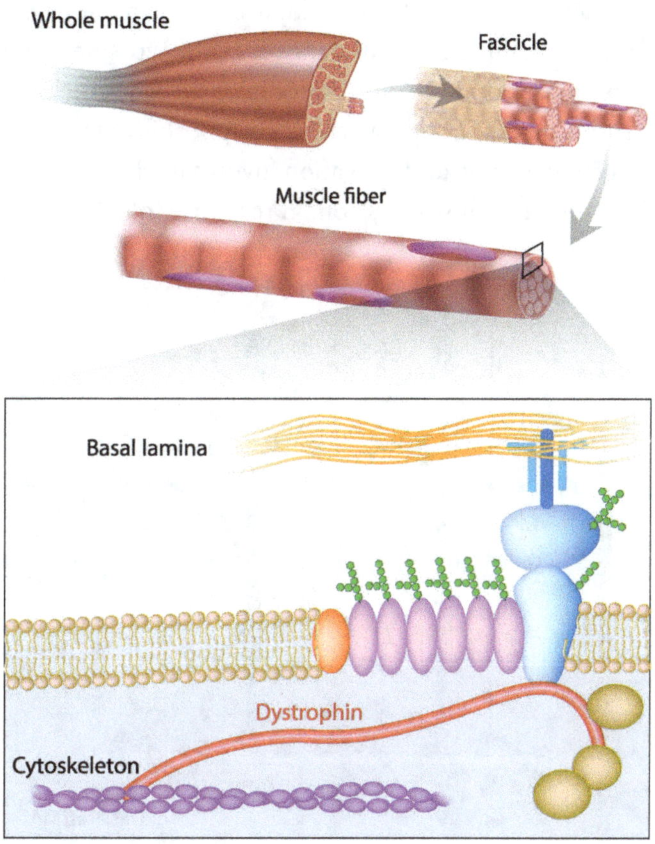

Alila Medical Media/Shutterstock.com

FIGURE 5-2. The structure and function of dystrophin

5.2.3 SIGNALING PROTEINS

These proteins perform the function of transferring a molecular signal from without the cell to the nucleus to activate gene expression, or from within the cell to without and other cells. One of the earliest signaling proteins characterized is the epidermal growth factor receptor (EGFR). Its length spans across the cell membrane to bring the outside chemical signal, in the form of the presence and increased concentrations of the epidermal growth factor (EGF), to the inside. This can happen during wound healing, as the wounded tissue permits the translocation of the EGF to the side of the cell where the EGFR is located. The engagement of two EGFR molecules in an inactive dimer with the EGF in the receptor extracellular domain (ECD) induces a structural change, called a conformational change, to activate the tyrosine kinase domains (TKD) in their intracellular tails. This class of growth factor receptors are, therefore, called receptor tyrosine kinases (RTKs) (**Figure 5-3**), which made a cameo of an honorable mention in an episode of the TV show *"ER."* A distinguished-looking gentleman was introduced on the show as "the person who (molecularly) cloned the first receptor tyrosine kinase," which would have been the complementary DNA (cDNA) and gene coding for the EGFR.

The activated EGFR tyrosine kinase, by the namesake, phosphorylates the intracellular tails in a reciprocal manner in the dimer over many tyrosine residues. These phosphotyrosine residues (pTyr) serve as docking sites for intracellular proteins, including various kinases, to achieve a surge of concentration and trigger a cascade of phosphorylation and activation events. The final stages of the phosphorylation and activation involve nuclear translocation of gene expression activators and mobilization of gene expression, in response to the initial surge of EGF concentration outside the cell.

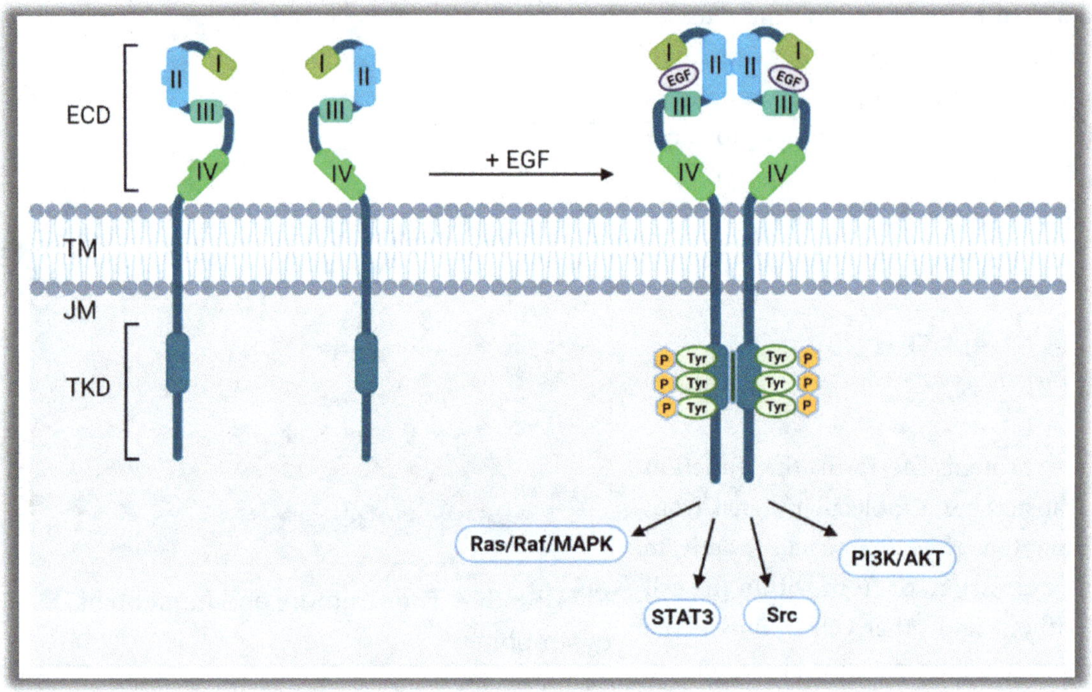

FIGURE 5-3. The structure and function of a receptor tyrosine kinase (RTK), such as the epidermal growth factor receptor (EGFR)

5.2.4 TRANSCRIPTION FACTORS

As their name would suggest, these proteins look for and identify the genes to activate in response to a signal, such as activation of the EGFR by the EGF for cell division in wound repair. The genetic text information contained in the linear sequence of a single-stranded nucleic acid, such as the DNA template for replication or transcription, can be "read" by base pairing with the aid of a polymerase. But the genetic text information in the intact genomic DNA is contained in the double-stranded form. The text letters, that is, the nucleobases, are not completely accessible to other molecules and would not permit base pairing as a mechanism of "reading" or "recognizing."

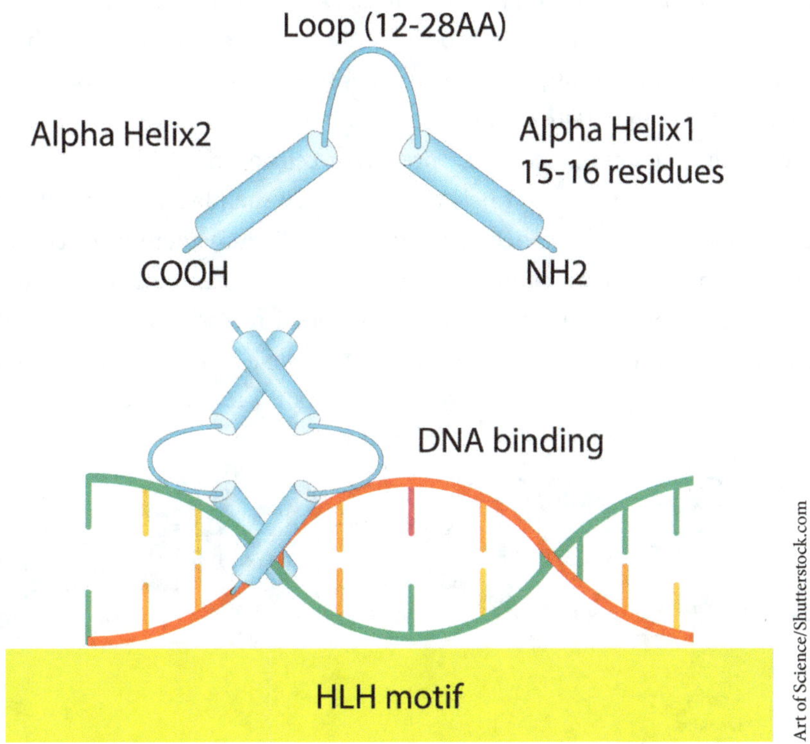

FIGURE 5-4. The structure and function of the helix-loop-helix domain in a transcription factor, interacting with double-stranded DNA

One prominent feature of the Watson-Crick double-helical structure of DNA was the illustration of a major groove and a minor groove spiraling along the length. In the major groove, the edges and corners of the nucleobases are characteristic of the specific base pairs, in all four different combinations, A-T, T-A, G-C, or C-G. Proteins with parts (domains) that can fit into the major groove, such as a helix-loop-helix (**Figure 5-4**) in the *c-myc* oncoprotein or a zinc finger domain in the Wilms tumor 1 suppressor protein (WT1), can achieve a high level of contact and form a stable complex with the double-stranded DNA at their specific recognition sequences. The engagement of different transcription factors with the DNA would result in a distortion of the perfect double-helical structure, either to open the DNA to become more accessible for transcription (activators) or to prevent the opening to silence transcription (repressors).

5.2.5 TRANSPORTER PROTEINS

Our body is a highly sophisticated and orchestrated multicellular and multi-tissue machine and must maintain a state of homeostasis. Inorganic chemicals cannot be allowed to freely diffuse across cells, tissues, and organ barriers and must be transported in a controlled manner to fulfill the local needs of the tissues or organs. The best example is the hemoglobin inside the red blood cells for the transport of oxygen. The rabbit β-globin cDNA and gene were the first to be cloned in a mammalian species, owing to its overwhelming abundance in the precursor cells of the red blood cells, the reticulocytes. Another good example are the ion channels across the cell membrane to maintain the proper electrical potential and osmotic fluid balance. The cystic fibrosis transmembrane conductance regulator (CFTR) protein is a chloride channel and is critical to the maintenance of fluid balance in the lung (**Figure 5-5**). Cystic fibrosis (CF) was the first autosomal recessive human hereditary disease for which the gene (*CFTR*) was mapped and identified by "reverse genetics" (linkage mapping).

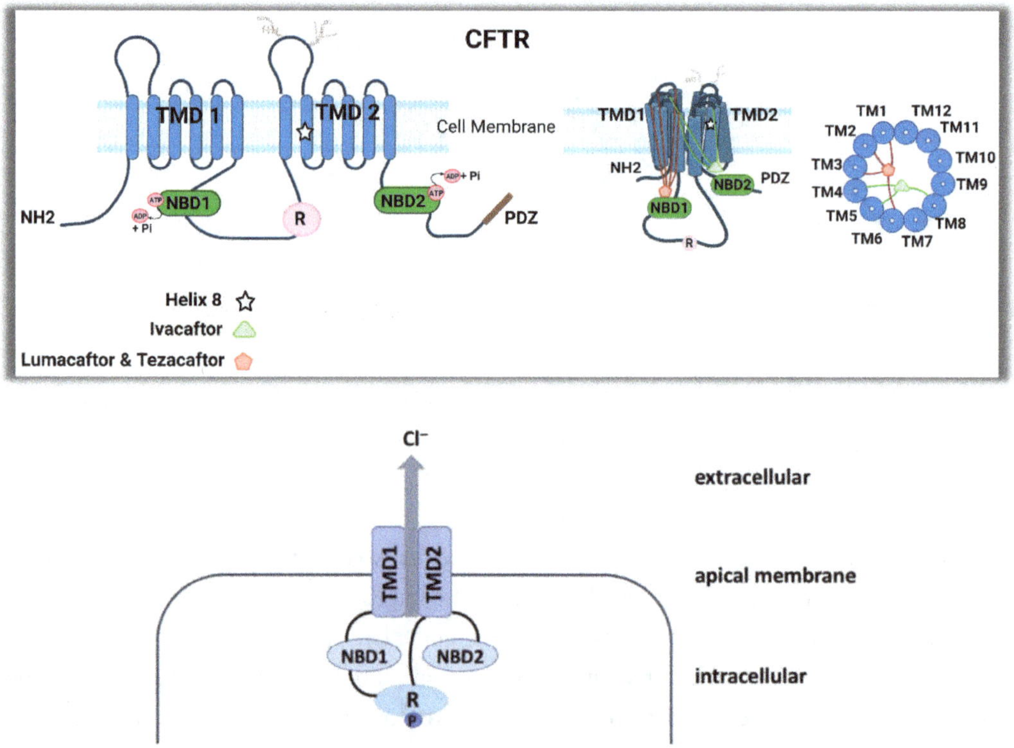

FIGURE 5-5. The structure and function of the cystic fibrosis transmembrane conductance regulator (CFTR)

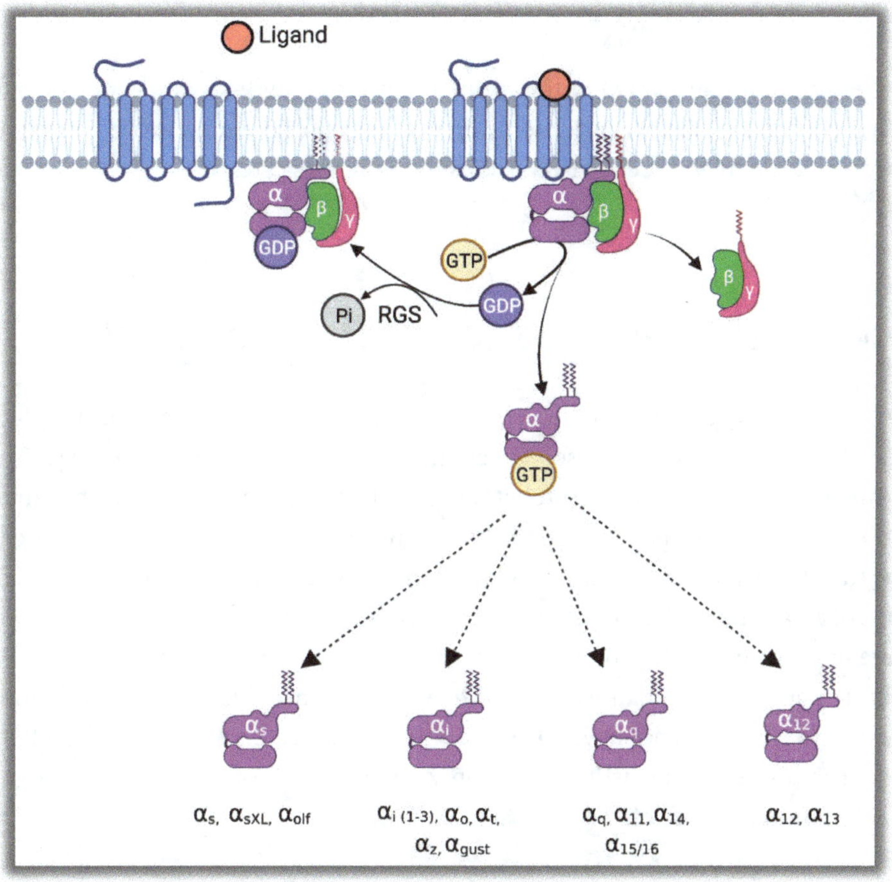

FIGURE 5-6. The structure and function of G-protein–coupled receptors (GPCRs)

5.2.6 G-PROTEIN–COUPLED RECEPTORS

These receptors are the largest family of proteins in humans, and targets of about half of all current drugs. They share a common feature in their protein structure that it passes through the cell membrane seven times and thus are also named seven-(pass)-transmembrane domain (7TM) receptors (**Figure 5-6**). The amino-terminal extracellular domain interacts with a wide variety of ligands, such as dopamine, epinephrine, histamine, serotonin, and many peptide hormones. The carboxyl-terminal intracellular domain couples to various subtypes of G proteins (guanine nucleotide–binding proteins) to trigger the downstream signal cascade. Our senses of vision, taste, and smell are controlled by members of the GPCR family.

5.2.7 IMMUNE AND DEFENSE PROTEINS

These include antibodies and small antimicrobial peptides (AMPs, such as defensins), as well as proteins that form the blood clots and intercellular junctions. The bulk protein albumin, the most abundant protein in the plasma, can also be considered in this category.

5.3 How Genetic Changes Affect Protein Function

The short summary above of protein structures and functions is not appropriate for that topic, but our focus here is on making "sense" of genetics, not on protein biochemistry. The key point is made by the two words often spoken together regarding proteins, structure and function, or the structure-function relationship. The proper biochemical word for the end product of the genetic information transfer is not a protein, it is a polypeptide of a chain of amino acid residues, that is, what remains of an amino acid in the chain. The mission of genetic information transfer is terminal and complete in the amino acid sequence of the polypeptide. The sequence, or order of the amino acid residues in a polypeptide chain is referred to as the "primary structure." It is a linear structure of sequence or order but does not have a three-dimensional shape. And so, if all proteins remain at the primary structure level, they will all have only one shape, a rope. And no protein would have any or much function to speak of.

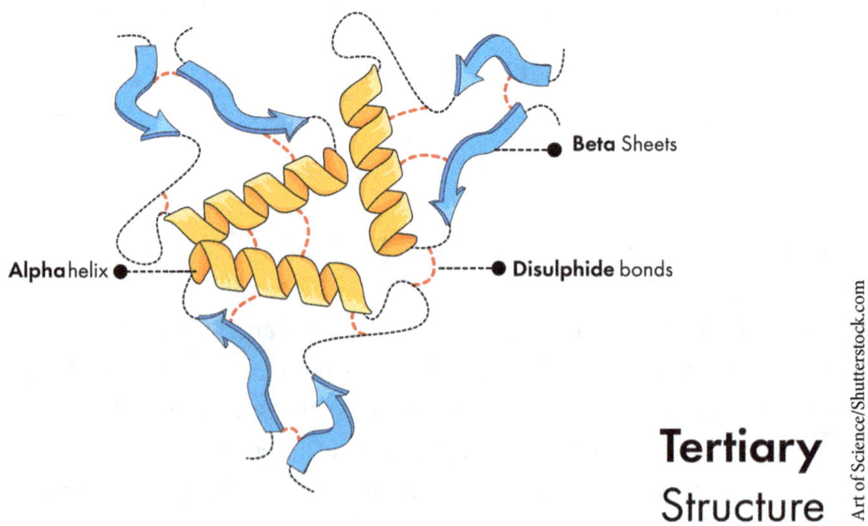

FIGURE 5-7. Protein folding with alpha-helices and beta-sheets into a tertiary structure

The American biochemist and peace activist Linus Pauling (1901–1994) had the distinctive honor of earning two unshared Nobel Prizes, one for his scientific work, the 1954 Nobel Prize in Chemistry, and the other for his peace activism, the 1962 Nobel Peace Prize. Pauling's Nobel Prize in Chemistry was awarded "for his research into the nature of the chemical bond and its application to the elucidation of the structure of complex substances." By studying the crystals of amino acids and peptides, Pauling described the planar structure of a peptide bond and proposed two basic "secondary structures" in a polypeptide (**Figure 5-7**), the α-helix and the β-sheet, in a 1951 paper in *PNAS*. These secondary structures form spontaneously upon the completion of synthesis of a polypeptide segment with a sequence conducive to the formation. They involve the carbonyl group of the peptide bond and the hydrogen of a nearby amine group in another peptide bond; thus, the stabilizing force is primarily hydrogen bond. Essentially, it is a backbone folding. Once the secondary structures are formed, the different parts of a polypeptide can interact with one another over the "long distance," involving additional hydrogen bonds, ionic interactions, and the force endowed by the tendency of the hydrophobic side chains to repel water, known as hydrophobic interactions.

The result of this long-distance folding is the unique three-dimensional tertiary structure of a natively produced protein, now a proper word to use. Some proteins gain their native biological functions upon completion of folding, such as the pancreatic ribonuclease A (RNase A). The secreted proteins enter the lumen of the endoplasmic reticulum (ER) by the first part of the polypeptide known as a signal peptide, which is cleaved following ER entry. The folding, packaging, transport, and secretion go through a specific intracellular vesicular network and itinerary. Some proteins would require activation by modification, such as the secreted blood coagulation factors, or engagement with a ligand, such as the intracellular steroid hormone (androgen and estrogen) receptors before assuming the final active form.

This again short introduction on the structure-function relationship of proteins gives us the basis to say that as small as a single amino acid change in a polypeptide may have the potential of altering the structure and in turn the function of a protein, if the change occurs in a part that is critical for the correct folding. Of course, there are changes that do not lead to such functional defects and thus tolerated. We can look at some specific examples from each class of protein described above.

5.3.1 ENZYMES—BLOOD CLOTTING FACTOR VIII

Hemophilia A is an X-linked recessive hemorrhagic disorder with an absence or decreased quantity or activity of the blood clotting factor VIII (FVIII), a serine protease required for the activation of thrombin, which releases fibrin from fibrinogen to form the blood clots (**Figure 5-1**). The clinical manifestation of hemophilia A is prolonged and excessive bleeding either spontaneously or after minor tissue damage. It is a prevalent genetic disease affecting 1 in 5,000 males, or more than 400,000 males in the world, accounting for 80% of hemophilia cases. The human FVIII gene is located on the X chromosome at about 186,000 base pairs (bp) in length, with 26 exons and an mRNA length of about 8,000 nucleotides. The mature FVIII protein is 2,332 amino acid residues long. The size of the gene and the length of its total coding sequence are comparatively speaking very large in the human genes; thus, the chances of mutations would be relatively high within this gene. Of more than 2,500 mutations identified, most are point mutations, that is, a change of one single letter in the DNA gene sequence. The majority are missense mutations (48.8%), that is, resulting in the change of a codon, thus an amino acid residue in the protein. The remaining are nonsense mutations (11.4%), that is, resulting in a termination codon that produces a truncated protein, and frameshift mutations (23.5%), that is, resulting in the wrong reading of the genetic code. Other mutations include joining errors of exons (splice site change), structural changes such as insertions and deletions, and synonymous changes, that is, not resulting in a change of the encoded amino acid, owing to the codon degeneracy.

5.3.2 STRUCTURAL PROTEINS—DYSTROPHIN

The best known example is dystrophin. The dystrophin gene is the largest in humans, at ~2.4 million base pairs (Mbp) in length with 79 exons. This is roughly 12 times larger than the FVIII gene, and 100 times the median human gene size at about 24 kbp (kilo base pairs). It is also about 8 times more complex than the average human gene with about nine exons. The enormous size and complexity present a tremendous burden to the cellular machinery that carries out the tasks of genetic information transfer, such as DNA replication and gene transcription. One of three DMD cases is caused by a new mutation, for example, a healthy father with the normal dystrophin gene generates a mutation in the reproductive cells (sperm) and passes it to a daughter. The normal X chromosome in the daughter (from her mother) would protect her from manifesting the disorder, but there is a 50% chance of her son to receive the X chromosome with the new mutation. Alternatively, a mother could also generate new mutations in one of the X chromosomes and pass it on to a child at a 50% chance.

Most of the dystrophin gene mutations are deletions (68%) or duplications (11%) of exons, whereas the remaining (20%) are small mutations. The deletions and duplications tend to occur in two regions, between exons 45-55 (deletions) and 2-10 (duplications). Deletions that do not affect the reading frame in the eventual mRNA sequence produce a dystrophin protein with two proper ends but shorter in the middle. The two ends of the dystrophin protein connect the cytoskeleton to the extracellular matrix, and the middle is the molecular shock absorber spring. The dystrophin with the shorter molecular shock absorber spring retains some function, and its clinical manifestation is milder, known as Becker muscular dystrophy (BMD). Deletions or duplications that result in a reading frame change would inevitably create an early termination, producing a truncated protein with major functional defects that is unstable and degraded intracellularly. The complete loss of function is manifested clinically as the more severe DMD. Point mutations and smaller deletions and insertions can also result in reading frame changes, early termination, or errors in joining the exons.

5.3.3 SIGNALING PROTEINS—EPIDERMAL GROWTH FACTOR RECEPTOR

Cell signaling plays pivotal roles in the embryonic development, and defects in the signaling proteins would often lead to embryonic lethality. The EGFR is a good example of such a signaling molecule. A complete loss of EGFR function, a condition comparable to the homozygote of an autosomal recessive disorder, can be created in the mouse using gene targeting to knock out both *Egfr* alleles. The litters die from embryonic days 12 to 15 (E12-15) to postnatal day 20 (P20) depending on the strain, displaying defects in the placenta, brain, bone, skin, and lung. No loss-of-function germline mutations have been reported for the human *EGFR* gene.

In the RTK family of proteins, only a small number of germline loss-of-function mutations has been recorded in humans with embryos developing to term. There is a germline point mutation G→A in the exon splice site of intron 10 in the *ErbB3* gene, closely related to the *EGFR/ErbB1* gene. This mutation results in a mistake in the splicing and joining of exons 10 and 11 and an insertion of eight nucleotides in the mRNA with the creation of a termination codon. The early termination creates a truncated ErbB3 protein at only about one-third of the normal total length. Clinically, the manifestation is an autosomal recessive disorder known as "lethal congenital contracture syndrome-2" (LCCS2), presenting severe multiple congenital contractures with muscle wasting and atrophy. Most patients die in the neonatal period with malformation of multiple organs, but two cases of survival to early adolescence have been reported. In mice, the *ErbB3* gene is shown to be expressed in the endocardial cushion mesenchyme in the developing heart to form the heart valves. Mice with homozygous loss of the *ErbB3* gene via gene targeting fail to completely develop the heart valves and die on embryonic day 13.5 (E13.5).

On the other hand, the EGFR is well known for its gain-of-function mutations in the somatic cells, and these mutant EGFRs are molecular targets for treating cancers driven by the mutant growth factor receptor. The small molecule tyrosine kinase inhibitor (TKI) gefitinib (Iressa) is used to target EGFR variants caused by point mutations in the *EGFR* gene, such as S768I, L861Q, and G719X. A variant EGFR caused by an exon 20 mutation can be targeted by the monoclonal antibody Rybrevant (amivantamab). The T790M EGFR mutant is a germline variant found initially in the southeast United States and associated with high incidence of familial non–small cell lung cancer (NSCLC).

5.3.4 TRANSCRIPTION FACTORS—TWIST1

The Twist-related protein 1 (TWIST1) is a basic helix-loop-helix transcription factor involved in early embryonic development regulating the folding of the single-layered embryo (blastula) into a multilayered and multidimensional structure (gastrula), in the process known as gastrulation. TWIST1 continues to have a critical role in mesodermal development. In a very familiar pattern as we have seen above for proteins involved in embryonic development, a homozygote *Twist1* gene knockout in mice is lethal on E11.5. Mice with the loss of one *Twist1* allele (heterozygous) develop craniofacial and limb abnormalities, similar to the phenotypes of Saethre-Chotzen syndrome (SCS), an autosomal dominant disease in humans. Human patients present a range of malformations, such as short stature, craniosynostoses, high forehead, ptosis, small ears with prominent crus, and maxillary hypoplasia with a narrow and high palate. More than 100 mutations have been reported in the human *TWIST1* gene, including nonsense and missense mutations, small insertions or deletions, and large gene deletions.

In the mouse and the human conditions with the loss of one allele at the *Twist1*/TWIST1 gene locus, the defects are manifested. The condition would have been a recessive trait if the remaining normal allele, with its transcription and translation into a normal protein were able to support the required functions. This phenomenon may be caused by two molecular mechanisms. One is called haploinsufficiency, literally meaning half is not enough. Ordinarily, the two alleles/copies of a gene in the somatic cells of a diploid organism are both expressed, referred to as biallelic expression. If the quantitative output of gene expression at the protein end from a single allele, that is, monoallelic expression, is sufficient biochemically or biologically to support the required functions, the loss-of-function mutations would follow a recessive mode of inheritance. In haploinsufficiency, half of the production is not enough, and the effect of the mutated allele is manifested despite the presence of the normal allele. Another mechanism is called dominant negative, literally meaning a loss of function manifesting as a dominant trait instead of a recessive one. In this situation, the mutant protein product interferes with the function of the normal protein made from the normal allele, creating the dominant phenotype. In the case of TWIST1 molecular genetic pathology, either mechanism can be at play, depending on the specific mutation in a variant allele.

5.3.5 TRANSPORTER PROTEINS—CYSTIC FIBROSIS TRANSMEMBRANE CONDUCTANCE REGULATOR

CF is an autosomal recessive disease caused by mutations in the *CFTR* gene, affecting the upper and lower airways, the gastrointestinal and reproductive tracts, as well as the endocrine system. CF incidence is the most common among Caucasians, at 1 in 3,000 to 4,000 live births; and 1 in 25 to 30 are carriers of a pathogenic CFTR allele. For Latin Americans, the prevalence is 1 in 4,000 to 10,000, and for African Americans 1 in 15,000 to 20,000. CF is uncommon in Africa and Asia, the frequency in Japan is 1 in 350,000.

The CFTR is an efflux (outgoing) chloride channel presented at the apical membrane of epithelial cells, such as the airway epithelial cells, regulating chloride secretion and osmotic pressure to bring water from inside the cells out to the lumen to hydrate the airway epithelia, by the airway surface liquid (ASL). CFTR defects reduce chloride secretion and increase sodium absorption with water, leading to a reduced ASL volume and a mucus buildup. The thick and sticky mucus in patients with CF causes difficulties in breathing and frequent lung infections.

The human *CFTR* gene is located on chromosome 7q31 at about 250 kbp in length, with 27 exons that produce an mRNA of 6.5 kb. The length of the CFTR protein is 1,480 amino acid residues, with five structural domains: two membrane-spanning domains (MSDs or TMDs), two nucleotide-binding domains (NBDs), and a regulatory (R) domain (**Figure 5-5**). There are ~2,000 mutations reported in the *CFTR* gene with varying degrees of disease severity, depending on the types and locations of the mutations and the parts of the CFTR protein the changes affect. The most frequent CFTR mutation is ΔF508, a deletion of the phenylalanine (Phe/F) at position 508 in the protein sequence, accounting for 70% of the CF cases. This single amino acid residue deletion causes folding errors that trap the folding intermediates in the ER, reducing the requisite apical membrane presentation and chloride channel activity.

In a different type of transporter protein, the hemoglobin in the red blood cells transporting oxygen to tissue and organs, a single-point mutation in the β-globin gene results in the replacement of a Glu with a Val in the mature protein. The variant form polymerizes into rigid fibers upon deoxygenation, deforming the red blood cells into the sickle shape to cause capillary blood vessel blockage and hemolysis. This is the mutation in the autosomal recessive disorder sickle cell disease/anemia (SCD).

5.3.6 G-PROTEIN–COUPLED RECEPTORS—RHODOPSIN

The current number of protein-coding genes in the human genome is 19,062, of which about 800 code for G-protein–coupled receptors (GPCRs), roughly 4%, or 4 out of every 100 genes. Greater than 2,350 mutations in 55 *GPCR* genes have been associated with 66 different human genetic diseases, while loss-of-function mutations in another 91 *GPCR* genes have been found without identifiable phenotypes. The prototypic human *GPCR* gene and the first to be molecularly cloned and sequenced in 1984 is the rhodopsin (*RHO*) gene, after the cloning and sequencing of the bovine rhodopsin gene. The sequence conservation across the two species enabled the use of the bovine RHO gene as a probe for the cloning of its human counterpart. The bovine RHO gene was cloned via first cloning its cDNA using degenerate oligonucleotides "back translated" from the carboxyl-terminal amino acid sequence of the purified bovine RHO protein.

The intent behind the cloning of the human rhodopsin gene was to enable the study of genetic differences causing inherited variations in human color vision. The RHO locus was pursued as a candidate gene, based on hypothesized functions, for some inherited forms of retinitis pigmentosa (RP), a heterogeneous group of inherited ocular diseases of progressive retinal degeneration and vision loss, with a prevalence of 1 in 3,000 to 5,000. The RHO-related retinitis pigmentosa is referred to as RP4, caused by mutations (more than 180) in the *RHO* gene located on human chromosome 3q22 and is autosomal dominant.

5.3.7 IMMUNE AND DEFENSE PROTEINS—BRUTON TYROSINE KINASE

There is a broad group of human hereditary diseases collectively known as "inborn errors of immunity" (IEIs), manifested as immunodeficiency, autoinflammation, autoimmunity, allergy, and development of cancer. The IEIs were initially thought to be rare and monogenic disorders of the molecular players in the immune system and only affect children. In 2022, the International Union of Immunological Societies Expert Committee updated the classification of human IEIs to a total number of 485, including 55 novel monogenic gene defects discovered after the previous update in 2020. These diseases encompass the entire spectrum of molecular mechanisms, such as gene dosage and expression changes, and gain or loss of protein functions; as well as the modes of inheritance, such as autosomal dominant or recessive, and X linked.

The first monogenic IEI was identified in 1952 by Colonel Ogden Bruton (1908–2003), a pediatrician at the Walter Reed Army Hospital, from an 8-year-old boy patient with very frequent pneumonia. The disease is the recessive X-linked agammaglobulinemia (XLA), literally meaning lacking γ-globulin in the blood, due to the absence of B cells, which are responsible for producing antibodies. The main proteins that fall under the γ zone of electrophoretically separated serum proteins are the antibodies. Bruton made the initial discovery that the serum from his young patient lacked the γ zone of serum proteins in the electrophoretic analysis. The causative gene for XLA encodes the protein Bruton tyrosine kinase (BTK), named after the discoverer of this IEI.

A group of disorders with defects in the development or function of immune cells, both T cells and B cells, are called "severe combined immunodeficiency" (SCID). Without treatment, SCID is fatal within the first 2 years of life. Genetic linkage of SCID has been shown to more than a dozen genes with an autosomal recessive mode of inheritance. The best known example is the gene encoding the adenosine deaminase (ADA), an enzyme essential for T-cell survival. In 1990, W. French Anderson pioneered human gene therapy to treat a 4-year-old girl, Ashanthi DeSilva, with ADA deficiency and SCID by using a retroviral vector to deliver a normal copy of the *ADA* gene. The gene therapy was a stunning success after 2 years of follow-up treatments, with the patient regaining a normal immune system.

BIBLIOGRAPHY

1. Aartsma-Rus A, Ginjaar IB, Bushby K. The importance of genetic diagnosis for Duchenne muscular dystrophy. *Journal of Medical Genetics.* 2016 Mar;53(3):145–51.
2. Amaral P, Carbonell-Sala S, De La Vega FM, et al. The status of the human gene catalogue. *Nature.* 2023; 622:41–7.
3. Avery OT, Macleod CM, McCarty M. Studies on the chemical nature of the substance inducing transformation of pneumococcal types: Induction of transformation by a desoxyribonucleic acid fraction isolated from pneumococcus type III. *Journal of Experimental Medicine.* 1944 Feb 1;79(2):137–58.
4. Beadle GW, Tatum EL. Genetic control of biochemical reactions in neurospora. *Proceedings of the National Academy of Sciences of the United States of America.* 1941 Nov 15;27(11):499–506.
5. Chen RZ, Cheng X, Tan Y, Chang TC, Lv H, Jia Y. An ENU-induced mutation in Twist1 transactivation domain causes hindlimb polydactyly with complete penetrance and dominant-negatively impairs E2A-dependent transcription. *Scientific Reports.* 2020 Feb 12;10(1):2501.
6. Eisenberg D. The discovery of the alpha-helix and beta-sheet, the principal structural features of proteins. *Proceedings of the National Academy of Sciences of the United States of America.* 2003 Sep 30;100(20):11207–10.
7. Griffith F. The significance of Pneumococcal types. *Journal of Hygiene* (London). 1928 Jan;27(2):113–159.
8. Guillamat-Prats R. Role of mesenchymal stem/stromal cells in coagulation. *International Journal of Molecular Science.* 2022 Sep 8;23(18):10393.
9. Hanssens LS, Duchateau J, Casimir GJ. CFTR protein: Not just a chloride channel? *Cells.* 2021 Oct 22;10(11):2844.
10. Hershey AD, Chase M. Independent functions of viral protein and nucleic acid in growth of bacteriophage. *Journal of General Physiology.* 1952 May;36(1):39–56.
11. Huang Y, Meng T, Wang S, Zhang H, Mues G, Qin C, Feng JQ, D'Souza RN, Lu Y. Twist1- and Twist2-haploinsufficiency results in reduced bone formation. *PLoS One.* 2014 Jun 27;9(6):e99331.
12. Inusa BPD, Hsu LL, Kohli N, Patel A, Ominu-Evbota K, Anie KA, Atoyebi W. Sickle cell disease-genetics, pathophysiology, clinical presentation and treatment. *International Journal of Neonatal Screening.* 2019 May 7;5(2):20.
13. Kaur G, Verma SK, Singh D, Singh NK. Role of G-Proteins and GPCRs in Cardiovascular Pathologies. *Bioengineering* (Basel). 2023 Jan 6;10(1):76.
14. McDonell LM, Kernohan KD, Boycott KM, Sawyer SL. Receptor tyrosine kinase mutations in developmental syndromes and cancer: Two sides of the same coin. *Human Molecular Genetics.* 2015 Oct 15;24(R1):R60–6.

15. Nair S, Bonner JA, Bredel M. EGFR mutations in head and neck squamous cell carcinoma. *International Journal of Molecular Science*. 2022 Mar 30;23(7):3818.
16. Narkis G, Ofir R, Manor E, Landau D, Elbedour K, Birk OS. Lethal congenital contractural syndrome type 2 (LCCS2) is caused by a mutation in ERBB3 (Her3), a modulator of the phosphatidylinositol-3-kinase/Akt pathway. *American Journal of Human Genetics*. 2007 Sep;81(3):589–95.
17. Nathans J, Hogness DS. Isolation, sequence analysis, and intron-exon arrangement of the gene encoding bovine rhodopsin. *Cell*. 1983 Oct;34(3):807–14.
18. Nathans J, Hogness DS. Isolation and nucleotide sequence of the gene encoding human rhodopsin. *Proceedings of the National Academy of Sciences of the United States of America*. 1984 Aug;81(15):4851–5.
19. Oxnard GR, Chen R, Pharr JC, Koeller DR, et al. Germline EGFR mutations and familial lung cancer. *Journal of Clinical Oncology*. 2023 Dec 1;41(34):5274–84.
20. Ramananda Y, Naren AP, Arora K. Functional consequences of CFTR interactions in cystic fibrosis. *International Journal of Molecular Science*. 2024 Mar 16;25(6):3384.
21. Salen P, Babiker HM. Hemophilia A. [Updated 2023 Jul 17]. In: StatPearls [Internet]. Treasure Island, FL: StatPearls Publishing; 2024 Jan–. Available from: https://www.ncbi.nlm.nih.gov/books/NBK470265/
22. Schöneberg T, Liebscher I. Mutations in G protein-coupled receptors: Mechanisms, pathophysiology and potential therapeutic approaches. *Pharmacology Reviews*. 2021 Jan;73(1):89–119.
23. Song H, Liu Y, Tan Y, Zhang Y, Jin W, Chen L, Wu S, Yan J, Li J, Chen Z, Chen S, Wang K. Recurrent noncoding somatic and germline WT1 variants converge to disrupt MYB binding in acute promyelocytic leukemia. *Blood*. 2022 Sep 8;140(10):1132–1144.
24. Tangye SG, Al-Herz W, Bousfiha A, Cunningham-Rundles C, et al. Human inborn errors of immunity: 2022 Update on the classification from the International Union of Immunological Societies Expert Committee. *Journal of Clinical Immunology*. 2022 Oct;42(7):1473–507.
25. Tyagi R, Kumar S, Dalal A, Mohammed F, Mohanty M, Kaur P, Anand A. Repurposing pathogenic variants of DMD gene and its isoforms for DMD exon skipping intervention. *Current Genomics*. 2019 Nov;20(7):519–30.

6

Diploid to Haploid to Diploid—The Circle of Life

"Relativity does set limits on what humans can ultimately do. But the universe is not required to be in perfect harmony with human aspirations."—Carl Sagan

- "What will modern day humans evolve to?"
- "Why are mutations caused by radiation or chemicals always bad for us?"
- "Which came first, the chicken or the egg?"

These are some common questions all of us think of once every now and again, and we will take a genetics look at what the best answers may be.

6.1 Diploid to Haploid

To evaluate the good or the bad of a genetic, that is, DNA change in terms of its effect on the organism, we separate the multicellular organisms from the single cellular organisms. We understand that every chemical/biochemical reaction obeys essentially one rule that it proceeds in a passive manner. Chemistry textbooks and courses teach us about reaction constants, rates, equilibrium, and free energy. The calculations that we perform may lead us to believe that we are in control. We are, but only to the extent of if we want to raise the production of vodka, we would need to use more potatoes as the starting raw material.

6.1.1 FROM ONE CELL TO TENS OF TRILLIONS

Genetics began as a human curiosity about why sons look like their fathers, and up to this point we have focused on the passage of the genetic information via the genetic material DNA transgenerationally. The consideration of genetics in the context of passaging genetic information via DNA is equally important for all of our body's cells, known as somatic cells, from Greek, soma, meaning "body." Humans have an estimated 37–100 trillion cells, or 37,000,000,000,000–100,000,000,000,000, in numbers based on various studies. We can use a middle number, 70 trillion. All of them came from one starting cell, the fertilized egg, containing the genetic material from both the sperm and the egg.

Humans are not the most efficient animal at reproduction, with a 25%–30% success rate from the fertilized egg. The giant panda is notorious for reproductive difficulties, but their major problem lies with the 2-year reproductive cycle of the female. The birth success rates for the giant panda, which is actually a bear, are 60.7% for natural mating, 50.6% for combined natural mating and artificial insemination techniques, and 18.5% for artificial insemination. The original panda, the red panda, has a birth success rate of 60%, as well. Both outperform the humans. Given the limited scope of our focus here, we will not address all the factors that make human reproduction rather difficult and inefficient. We will focus on the challenges in genetics, that is, the transfer of genetic information along the axis of the central dogma from the first cell of a fertilized egg, to the 70 trillion in a fully grown adult, and on, transgenerationally.

6.1.2 THE MITOTIC CELL DIVISION

The brand of genetics that we discuss here, especially with regard to humans, is "diploid multicellular organism genetics." The 70 trillion cells in a human body, including the cells that give rise to the egg and the sperm, are diploid, meaning that each cell contains a pair of each numbered chromosome, from 1 to 22, and the sex chromosomes, XX or XY. The process by which one diploid somatic cell becomes two is a cell division called mitosis, with the exact duplication of all chromosomal material, including the genomic DNA and the proteins (histones and nonhistones) before the cell division. Mitosis was first used in 1882 by the German anatomist Walther Flemming (1843–1905) from the Greek word "mitos," meaning "warp thread," quite appropriate for the appearance of the chromosomes observed during cell division. The DNA replication and chromosome duplication occur in the S (synthesis) phase of the cell cycle. The duplicated chromosomes are called chromatids, a term suggested in 1900 by Clarence Erwin McClung (1870–1946). They are linked physically at the centromere kinetochore.

The key molecular mechanism controlling the proper distribution of chromosomes in the "daughter cells" after cell division is the "cohesion" of the duplicated but still bound together sister chromatids. Cohesion is performed by a protein complex called cohesin, which locks the 10-nm or 30-nm DNA-core histone fibers across the two sister chromatids along the length. To prepare for the proper sorting of the duplicated chromatids to the equatorial plate of the cell for a clean separation during division, two additional major events occur during the M phase (metaphase) of the cell cycle. The long chromatids are "condensed," by a protein complex called condensin, similar to the cohesin in structure, locking the 10-nm or 30-nm DNA-core histone fibers within each chromatid along the length. The condensed chromatids are fatter but also much shorter, and easier to move and sort. The familiar images of chromosomes we see in books are photographed condensed metaphase chromosomes (without cohesins) after staining, typically with Giemsa staining to produce the G-bands, as shown in **Figure 6-1** by the ideograms. Giemsa staining of chromosomes was invented by the German chemist and bacteriologist Gustav Giemsa (1867–1948).

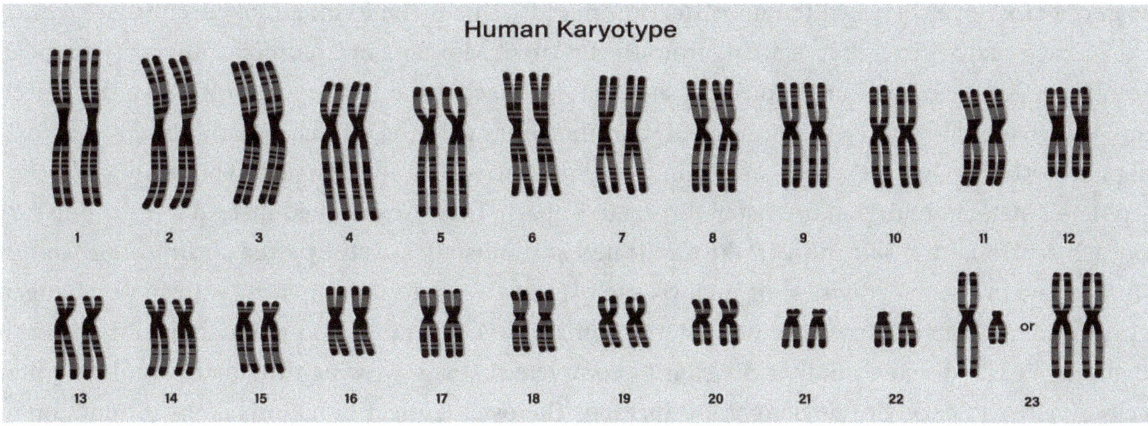

Dee-sign/Shutterstock.com

FIGURE 6-1. The human karyotype displaying the numbered 22 autosomes and the X/Y sex chromosomes

Moving and sorting are done by the spindle fibers (strings) attached to the kinetochores while the cohesins hold the sister chromatids together. Once all the molecular and cellular milestones have been achieved the cohesins are unlocked by a protease (aptly named separase) and the sister chromatids are pulled apart and evenly distributed in the two daughter cells. We used "milestones" to refer to all other molecular and cellular requirements for a successful cell division and continued cycles because we want to focus on and highlight the events concerning the chromosomes and the DNA therein.

6.1.3 THE MEIOTIC DIVISION—MORE THAN "LESSENING"

The division that produces the egg and the sperm is called meiosis, first used in 1905 by J. Bretland Farmer (1865–1944) and J.E.S. Moore (1870–1947), from Greek, meaning "lessening." The part being lessened in this type of cell division is the number of chromosomes. The meiotic division is not different from a mitotic division in terms of the overall molecular and cellular programs, except for some special key details. The most important difference is a component that makes up the cohesins during a meiotic division, due to the hormonally regulated environment where the precursor (pregametic) cells divide to produce the egg and the sperm. In these cells the regular cohesin component gene is not turned on but the special cohesin component gene is turned on. The meiotic cohesins stay longer on the sister chromatids and require an extra biochemical step before they can be unlocked by separase.

As a result, the homologous (same-numbered) chromosomes with their duplicated chromatids can "pair" together by forces that are largely unknown, and the pairing is stabilized by chromosomal material cross-over between the nonsister chromatids. This is the foundational event that gives us the linkage analysis exercises, starting from the fly lab of Morgan's in Columbia University to today in human genetics. The same molecular and cellular mechanical forces will now move the paired and locked homologous (same-numbered) chromosomes with the four chromatids to the equatorial plate for cell division, except that herein the spindle fibers are not attached to both sides of a centromere kinetochore to pull the sister chromatids apart. They are attached instead to only one side of each centromere kinetochore in the duplicated chromosomes, as the paired chromosome stands in the way. In the first division of meiosis each set of two sister chromatids, with the exchanged chromosomal material from the nonsister chromatids, is separated into two daughter cells. Each daughter cell undergoes another division to complete meiosis, in which the first-round daughter cells give rise to another two daughter cells each. The overall effect of meiosis is the production of four gamete cells, each with one haploid slate of chromosomes (**Figure 1-3**).

In men, one pregametic cell produces four sperm, and in women there is one mature egg cell and the other three are polar bodies and reproductively irrelevant. In men, meiosis starts with the pregametic cell and proceeds through. But in women the first meiotic division of all pregametic cells progresses to the G2 phase, in between the S and the M phases where DNA replication quality control and damage repair are performed. They are then "arrested" in the G2 phase in the embryonic stage. After sexual maturity one egg matures *per* month in an adult woman, whereas a healthy adult man produces hundreds of millions of sperm *per* ejaculation. In theory, the huge differences in the number of gametes produced by the male *versus* the female should not affect the probabilities of allele–allele recombination across two loci of interest. But in effect, due to the physiological and cellular environment differences in the reproductive organs, the chromosome (chromatin) structures are different between the male and the female. Overall, the female chromosomes have more crossover events and their chromosomes appear "bigger" when measured by genetic linkage distances than the male counterparts, despite the identical physical sizes of DNA in the same-numbered chromosomes.

6.2 Reproduction Requires Stemness

"Which came first, the chicken or the egg?" If we look at the cellular machinery involved in cell division, the most important difference between a mitotic and a meiotic division is the different cohesin complexes being made in the somatic *versus* the reproductive cells. This molecular difference with a big cellular behavior impact should argue for the chicken to have come first. But that would require for the chicken to be able to reproduce asexually, that is, from diploid to diploid, before the egg came about. Can multicellular organisms reproduce asexually?

6.2.1 DOLLY THE SHEEP WAS AN EXAMPLE OF ASEXUAL REPRODUCTION

The sensational story of Dolly the Sheep in 1996 was an example of animal cloning, or, asexual reproduction. From the NHGRI: "The term cloning describes a number of different processes that can be used to produce genetically identical copies of a biological entity. The copied material, which has the same genetic makeup as the original, is referred to as a clone. Researchers have cloned a wide range of biological materials, including genes, cells, tissues and even entire organisms, such as a sheep." Animal cloning starts with taking the nucleus from a somatic cell, containing the diploid chromosomes and the genomic DNA therein and

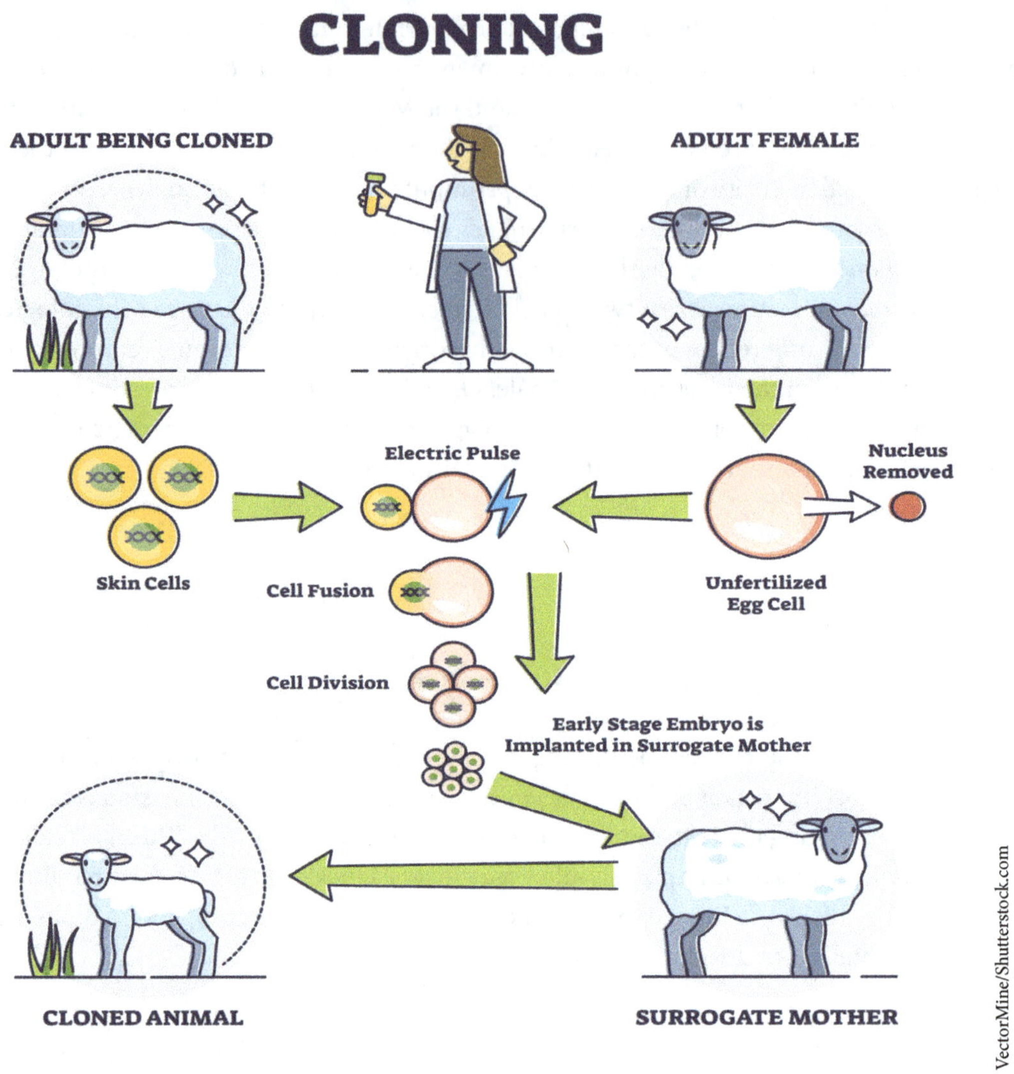

FIGURE 6-2. Animal cloning is asexual reproduction

transferring it into an anuclear egg cell, that is, having had its own nucleus removed (**Figure 6-2**). The egg with the somatic cell nucleus is grown *in vitro* into an early embryo before implantation in a foster mother to carry to term and birth. Dolly's nuclear donor cell was from the udder epithelium of a six-year-old sheep, thus the name Dolly (after Parton). The often untold part of the story in the public is that Dolly was one success after 276 attempts. Other animals have been cloned to date, including cats, dogs, horses, rabbits, and rhesus monkeys. But the success rates remain low and the procedures complex and expensive.

These organisms do not naturally reproduce asexually because the trillions of cells in the adults are highly specialized, known as differentiated, literally, being made different. How they are different and what made them different are very large topics so we look at some key aspects, again with a focus on the transfer of genetic information along the axis of the central dogma. Multicellular complex diploid organisms such as humans reproduce via a haploid phase involving the gametic cells of two sexes. We can think of this as a "bypass" of differentiation.

6.2.2 THE MOLECULAR GENETIC DIFFERENCES IN OUR CELLS—CHAPTERS OF THE BOOK OF LIFE

The sperm and the egg contain the purest form of genetic information in their genomic DNA with all epigenetic marks erased during gametogenesis. The 70 trillion cells in an adult body, male or female, can be described in more than 200 different histological types as they appear under a microscope. But their "molecular landscape," viewed from the functional output end of all the proteins in variety and quantitative variations, can be infinite in types when the cellular environment changes and the cells make different proteins in different quantities to adapt to the changes. Yet, all 70 trillion contain, by design, the same exact genomic information content, in the diploid genome of 3 billion base pairs in each haploid set. Of course along the way of development and aging, the DNA in the cells pick up changes, replication errors, or chemically induced changes. For the current discussion, we do not need to factor in these changes and can consider the genomic DNA being the same in all 70 trillion cells. The confidence that we have for doing this is based on the most recent estimates of mutation accumulation in humans, the substitution rates *per* genome (3 billion) *per* year is 47, so roughly 1 millionth of 1.57%.

Now, how do we have at least 200 different types of cells when examined under a microscope while all 70 trillion cells use the same genetic information textbook content? The answer is, chapters and highlights. In each cell type, for instance the liver cells, only the proteins required for liver development and functions are produced by transcribing the genes encoding them from the specific parts of the genome. In the intestinal epithelial cells, those that line the inside of the gut, a different set of proteins from their corresponding genes are produced. The skeletal muscle cells do quite different things from the first two and would make their own cell-type-specific set of proteins using the corresponding genes. Some proteins are produced in all three and all other cell types. These are called "housekeeping genes (proteins)," as the basic structures and metabolic functions of a cell are common. In each cell type, the varieties and quantities of each protein would also change when the cells are given a stimulus, for example, liver cells upon alcohol consumption, gut cells during an infection, and muscle cells after strenuous exercises. The activation of different genes in different cell types and under different conditions from essentially an identical gene book can be viewed as "highlighting" different parts of the text and collating the highlighted parts in different excerpt chapters.

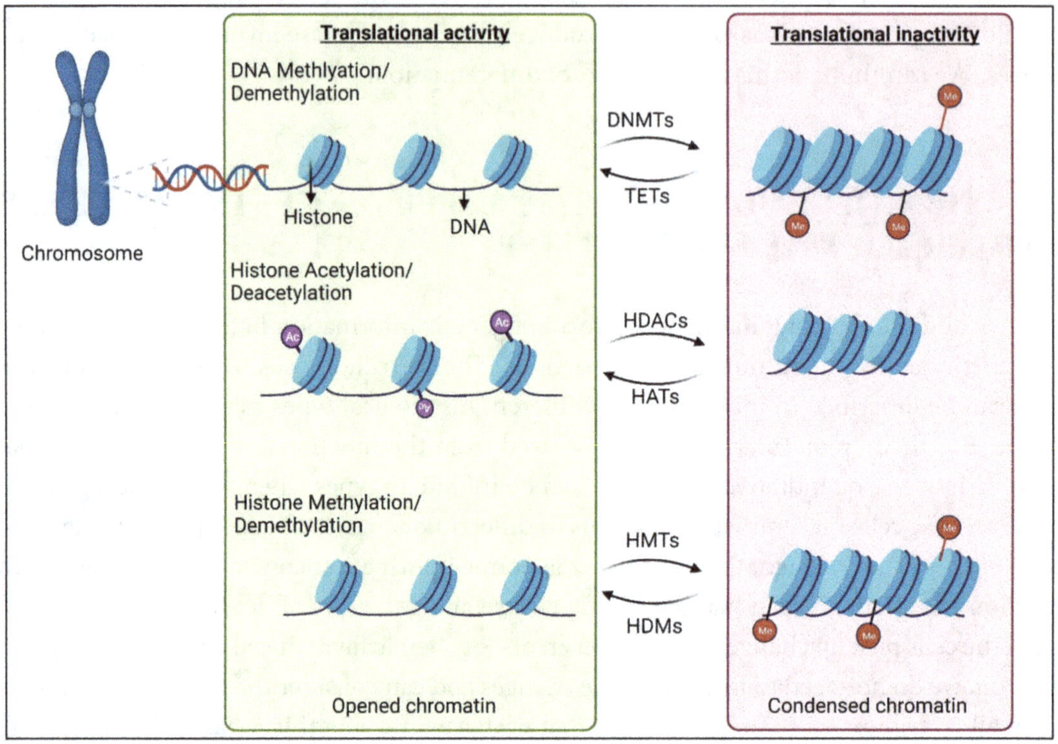

FIGURE 6-3. The chemical modifications of the DNA and histone tails control which regions (chapters) of the genome (book of life) are open for expression (being read)

DNMT: DNA methyltransferase; TET: ten-eleven translocation enzymes; HDAC: histone deacetylase; HAT: histone acetylase; HMT: histone methyl transferase; HDM: histone demethylase; Me: methyl group; Ac: acetyl group.

The DNA in our chromosomes is not "naked" and is spooled around a cluster of proteins called "core histones," as shown in **Figure 6-3**. The histone proteins have extended amino-terminal tails that wrap around the spooled DNA, either tightly to restrict access to the DNA for gene expression or in a more relaxed way to allow access and gene expression. The "highlighting" is carried out chemically by modifications of the nucleobases in the DNA, the best understood is the methylation of cytosine on the 5-position of its ring, like adding a molecular tag, the methyl group (Me, $-CH_3$). The cytosine methylation occurs on the cytosine in a CpG dinucleotide, "p" denotes the backbone phosphate. In a double-stranded DNA, the two strands have an antiparallel, up-and-down inverted symmetry at a CpG dinucleotide. The DNA is considered fully methylated when both CpG cytosines are methylated. When only the cytosine in one but not the other strand is methylated, it is referred to as "hemi-methylated." DNA methylation in the regulatory regions of a gene tends to silence gene expression, whereas unmethylated DNA is more conducive to gene expression. Gene expression regulation also involves modifications of the histone tails, the best understood are methylation (Me) and acetylation (Ac), to restrict or allow DNA access.

6.2.3 MANY LEVELS OF STEMNESS

If an engineer instead of a biologist is asked: "What is the best way to dedifferentiate a cell from a terminally differentiated state?", the answer would probably be: "Keep it undifferentiated." The term we use to describe the undifferentiated cells is stemness, and the cells are referred to as stem cells, which are self-renewing. Terminally differentiated cells are incapable of self-renewal, unless manipulated. There are different levels or states of stemness. A totipotent stem cell is capable of developing into an entire animal, whereas a pluripotent stem cell is capable of developing into many different kinds of cells of the three different germ layers in a body. The germ layers—the endoderm, ectoderm, and mesoderm—are formed when the early developing embryo folds in the process known as gastrulation. Each layer gives rise to a specific group of cell types that form different tissues and organs. The innermost endoderm develops into the internal organs, such as the gastrointestinal track, lungs, and liver. The outermost ectoderm develops into the barriers of the body, such as the epidermis, but also the mammary gland and the nervous systems. The mesoderm in the middle gives rise to the dermis, heart, muscles, bone and bone marrow, and the urogenital system. A multipotent stem cell is capable of developing into different kinds of cell within one germ layer. A further narrower type of stem cells is the unipotent, capable of developing into a limited number of cell types. The cell that gave rise to Dolly the Sheep would have to be manipulated or "taught" to revert from a terminally differentiated mammary epithelial cell or a suspected mammary unipotent stem cell all the way back to a totipotent stem cell. At the chromosomal or genomic DNA level the reversion involved wiping the DNA and histone modification marks.

The reproductive cells came from an early embryonic lineage called the "primordial germ cells" (PGCs), which are pluripotent stem cells. There are reproductive physiology differences between the sexes in the maintenance of stem cells that produce the eventual gametes, the egg and the sperm. In the female, some PGCs undergo meiosis in the embryo to produce oocytes (several million) with the division arrested in the G2 phase. At birth there are around 1 million oocytes and typically one mature egg is produced with the completion of meiosis *per* cycle in the reproductive years of a woman. The remaining PGCs become the female germline stem cells (FGSCs), a rather recent discovery at the turn of the century. In the male, PGCs become gonocytes in the embryo and remain dormant, but resume division and differentiate into spermatogonial stem cells (SSCs) after birth. Meiosis in the male reproductive system begins at puberty to produce spermatozoa from the SSCs.

6.2.4 EPIGENETIC REPROGRAMMING FOR A CLEAN SLATE

In the differentiation of PGCs to become the gametic cells, there is one round of global erasure of the parental epigenetic DNA methylation marks down to approximately 5%, followed by *de novo* DNA methylation to establish germ-cell-specific marks differentially in the sperm and the egg (**Figure 6-4**). In the male (upward solid blue line), methylation in the genomic DNA completes before birth and before the PGCs enter meiotic divisions. In the female (upward solid pink line), this occurs after birth during maturation of the egg cells that complete the arrested meiotic divisions.

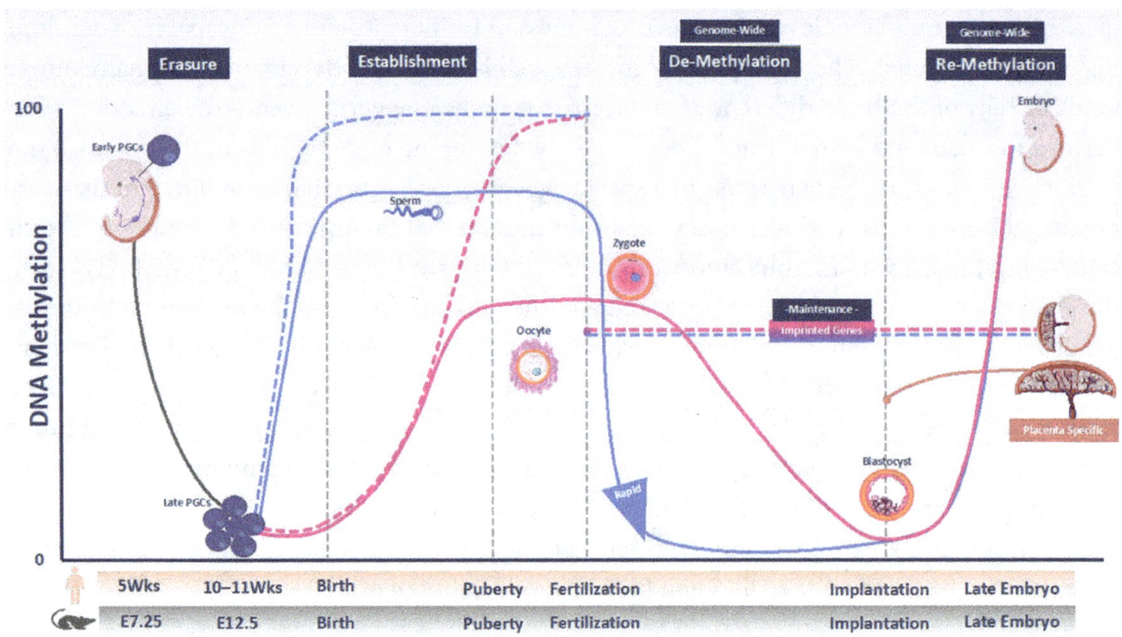

FIGURE 6-4. Epigenetic reprogramming in germ cells and early development

The global erasure is required to remove the differentially methylated regions (DMRs) specific for each parental genome (**Figure 6-5**), the epigenetic basis of imprinting. The two different copies of genomic DNA become essentially a clean slate for establishing the sperm- or oocyte-specific *de novo* methylation. In each sex, the male- or female-specific DMRs are also reestablished in this process, shown in **Figure 6-5**, and also by the dotted lines in **Figure 6-4**. These are maintained until the PGCs in the new life begin producing the gametes, in the next cycle. Now, the two different parental copies of the genomic DNA in the PGCs in either sex generate one single kind of parent-of-origin gametic DNA.

The irony of PGCs being set aside to not become any other terminally differentiated cells is that they are terminally differentiated into the gametic cells. In these modern days of molecular genetics, this means that they have germ-cell-specific patterns of gene expression and molecular landscape. To ensure totipotency, as the zygote will give rise to all cells in the body, the sperm and egg genomic DNA copies undergo another round of genome-wide demethylation (**Figures 6-4** and **6-5**) immediately after fertilization and before implantation. The paternal sperm genome is demethylated rapidly by an enzymatically catalyzed active process (**Figure 6-6**) before the first cell division from the fertilized egg (**Figure 6-4**, downward solid blue line). The maternal egg genome is demethylated gradually by a passive mechanism (**Figure 6-6**) through rounds of DNA replication during the preimplantation developmental period (**Figure 6-4**, downward solid pink line). The nucleotide substrates used for replication by the DNA polymerase are not methylated thus the newly synthesized DNA strand is not methylated.

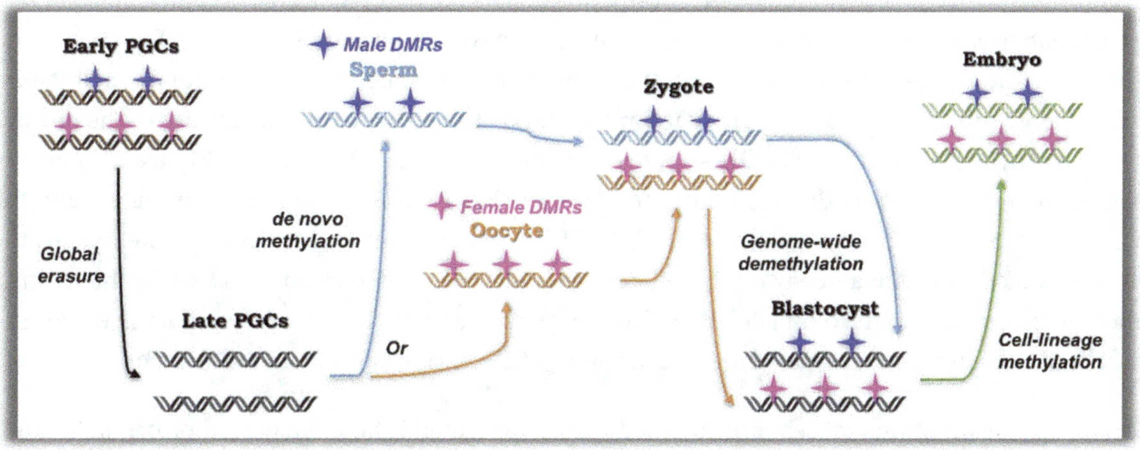

FIGURE 6-5. Erasure of pregametic DNA methylation and establishment of gametic methylation with differentially methylated regions (DMRs) in the two sexes, and the erasure of gametic methylation with maintenance of the DMRs in the zygote and embryos

The double helix image was created by "mcmurryjulie": https://pixabay.com/vectors/dna-dna-icon-icon-2316536/ (CC0), and used with modifications for **Figure 6-5**.

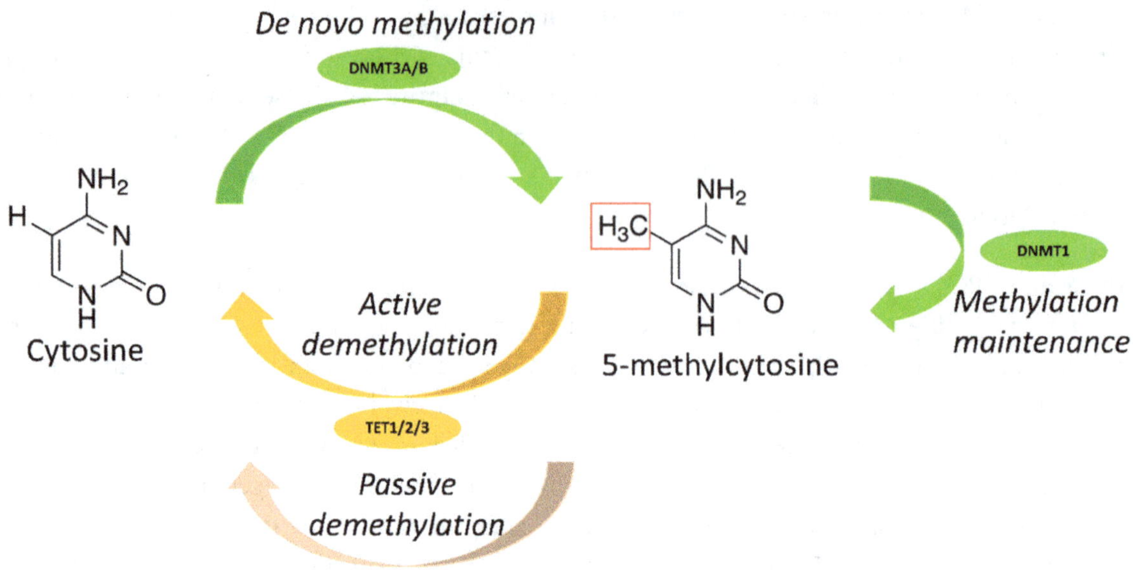

FIGURE 6-6. DNA methylation and demethylation

After implantation, the genomic DNA is re-methylated *de novo* to establish the embryonic and placental patterns of epigenetic marks (the merged upward solid blue and pink lines in **Figure 6-4** and the green line in **Figure 6-5**). Throughout the embryo and continued human development, different cell-type-specific patterns are acquired and maintained. The presence and quantities of the *de novo* DNA methylation enzymes, the DNA methyl transferases DNMT3A and 3B (**Figure 6-6**) and of the maintenance DNA methylation enzyme DNMT1 (**Figure 6-6**) cycle in correspondence to the waves of demethylation and methylation. DNMT1, when present and functional, methylates the unmethylated strand in a newly replicated hemi-methylated double-stranded DNA by the existing methylation marks on the template DNA. The maintenance methylation ensures that the epigenetic marks of cell lineages are retained after each round of DNA replication and cell division.

An intriguing question that remains to be actively investigated is how the parent-of-origin genomic imprinted regions resist the genome-wide demethylation upon fertilization and escape the *de novo* genomic methylation. The DNA is a passive biological macromolecule and by itself unable to respond differently to the enzymes that actively demethylate or methylate, or to the replication-dependent passive demethylation. The current hypothesis is that protein factors specifically protect these areas of DNA from unintended demethylation or methylation.

6.3 Haploid to Diploid

6.3.1 GENOMIC IMPRINTING—THE IMPORTANCE OF BOTH PARENTS

How important is it for the two different copies of the genome to maintain a certain level of imprinted marks from the parents of origin? In 1984, two laboratories led by Davor Solter and Azim Surani, respectively, performed nuclear transfer experiments in mouse embryos. Following fertilization, the nucleus of the sperm or the egg was removed and then added to create zygotes with either two maternal pronuclei or two paternal pronuclei, containing two copies of the genome from a single parent. Neither of these types of zygotes survived in the embryonic development. Fertilized eggs with two copies of the genome from a single parent also occur spontaneously in humans, but are also unable to survive through embryonic development. These events are the results of faulty chromosome distribution during meiosis and is referred to as uniparental disomy (UPD) if it involved one chromosome or genome-wide UPD (GW-UPD) if it involved all chromosomes. GW-UPD is observed more frequently with the paternal genome origin, at up to 13 *per* 1,000 pregnancies, whereas only a few cases of maternal GW-UPD have been reported.

Most important, sex-specific DMRs lead to sex-specific expression from the chromosome of one but not the other parent, a form of mono-allelic expression. The best hypothesis to the parental origin-specific gene expression by imprinting is the "parental conflict theory" by David Haig. By this theory, the paternally expressed genes promote growth to gain advantage in the acquisition of nutrients in the placenta over the siblings, especially relevant in polygamous species. The maternally expressed genes, on the other hand, would restrict growth to counteract against the effects of the paternally expressed genes. In mammals, a pair of such counteracting genes are *IGF2* (insulin-like growth factor 2) and *IGF2R* (IGF2 receptor), paternally and maternally expressed, respectively. In species that lay eggs, such as the fish and the bird, and the mammals platypus and echidna, there is no differential imprinting and expression of the growth-promoting or -restricting genes, as the nutrients are evenly distributed. In the egg-laying mammal, the marsupials that grow in the mother's pouch, however, imprinting and mono-allelic expression are observed. The parental conflict theory leaves the door open on the explanation of imprinted and mono-allelically expressed genes that are not involved in growth regulation.

Reproduction via haploid gametic cells is often said to offer the organisms the opportunities for producing more allelic diversities due to the unique events of chromosome pairing and crossing over in the first meiotic division. It is true but it can only be viewed as a fringe benefit as the main benefit is the relatively shorter and simpler route back to totipotent stemness.

6.3.2 DNA REPLICATION AND CELL DIVISION FROM ONE TO TRILLIONS

The first diploid cell that gives rise to all 70 trillion in our body is very different with regard to how it replicates the DNA. The DNA replication occurs separately in the pronuclei from the sperm and the egg before the fusion to create the nucleus of the fertilized egg. Early embryo DNA replication and chromosomal events during cell division are highly error-prone, due to the challenges created by the epigenetic reprogramming and reactivation of gene expression. Much of the highly orchestrated quality control machinery has not been fully oiled. Consequently, the genetic material damage and imbalance in human pregnancies can be rather high, with an about 17% rate of mosaicism, that is, not all cells contain the exact same genetic material. Mosaicism can result in pregnancy failure or survive to birth and remain in a healthy individual depending on the cell fate or if the mosaicism is rescued.

The joining of the sperm and the egg can very much be compared to a young couple moving in to start a new life together after marriage. They come from two different family backgrounds and environments (the male and the female reproductive systems), bringing over their personalized life habits (gamete-specific methylation) and idiosyncrasies (the DMRs). The couple learn to live with each other by erasing the personalized habits and reestablishing new habits (embryonic and placental epigenetic marks) for the new life together, while maintaining an identity (DMRs).

6.4 Maintenance of Homeostasis in Trillions and Hundreds of Types of Cells

At the cellular level, the first division gives us two cells, and the second gives us four, and then the third gives eight. We do not continue as a ball of identical cells, and at the 8- to 16-cell stage of the embryonic development, compaction and polarization take place. The cells on the outside and the inside then become different, with the ensuring hollowing to become the blastocyst containing the epiblast, hypoblast, and trophectoderm tissues. After implantation, some embryonic epiblast cells form the amniotic cavity and gastrulation folds the embryo to generate the three germ layers with their distinctive cell lineages (**Figure 6-7**).

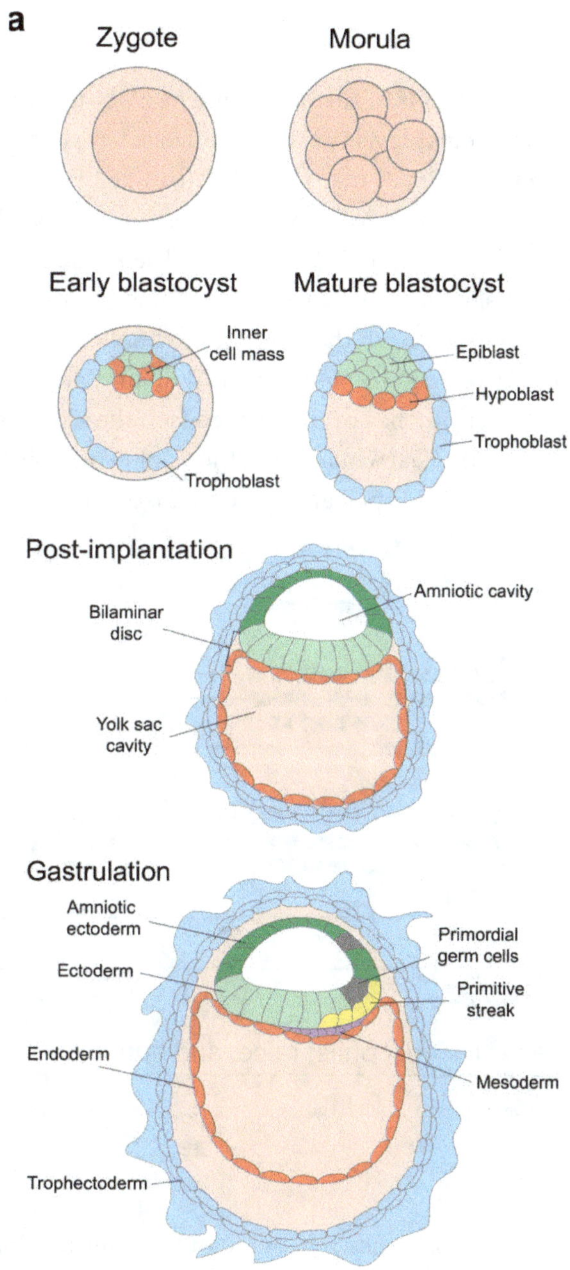

FIGURE 6-7. Early embryonic development of humans

A portion of the original image was used for Figure 6-7.

6.4.1 FATAL IN THE FLIES

We often speak of a "body plan" in terms of the development of tissues and organs to give the final form of the organism. At the molecular level, the best known set of genes that guide the formation of the proper body in a bilaterian animal, that is, with left-to-right symmetry are the homeobox genes. They share a characteristic "homeobox" DNA sequence of 180 bp, which codes for a 60-amino acid helix-turn-helix homeodomain for DNA binding. The homeobox proteins are thus transcription factors and have been shown to regulate the spatial and temporal patterns of tissue and organ development in animals ranging from the fruit flies to humans (**Figure 6-8**). Homeobox was named after the term "homeosis," coined by, once again, William Bateson, in 1894, to describe the variations that appear to be "something has been changed into the likeness of something else." His intent was to provide support for the Darwinian hypothesis of evolution with his work "Materials for the Study of Variation, Treated with Especial Regard to Discontinuity in the Origin of Species."

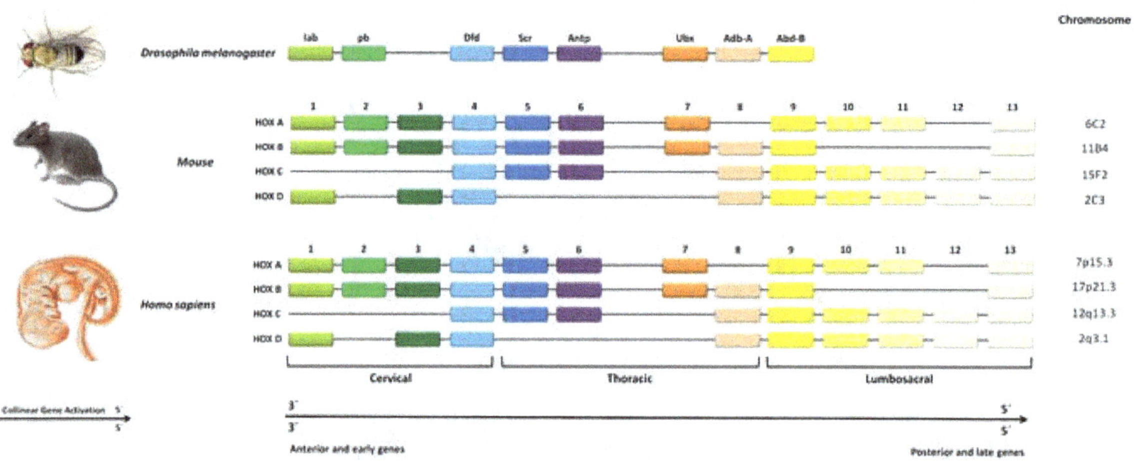

FIGURE 6-8. Homeobox genes in Drosophila, mice, and humans

The initial discovery of the homeobox genes was made in the fruit fly *Drosophila*, with mutations in the gene *Antennapedia (Antp)*. Mutations in the *Antp* gene that result in its loss of expression in the right place and at the right time, or abnormal expression in the wrong place and at the wrong time can force the second leg pair to develop into antennae or the antennae to develop into legs. The mouse *Hox-2.2/Hoxb6* gene is the counterpart of *Antp*, and when it is forced to be expressed in the flies by genetic engineering the same legs to antennae or antennae to legs phenotypes can be reproduced. In humans, the counterpart is the HOXB6 gene. The mammalian *HOXB6* (human) and

Hoxb6 (mouse) genes are involved in the differentiation of the erythroid progenitor cells. Knocking out the *Antp* gene in the flies is embryonic lethal, while mice with a double (homozygous) *Hoxb6* gene knockout are viable and fertile, but with functional defects in the erythroid progenitor cells. The *Hoxb6*-deficient mice had very specific skeletal abnormalities, such as the absence of the first pair of ribs associated with anterior transformations of the vertebrae in positions C6 through T3, mimicking features normally present in positions C5 through T2. No human disease has been associated with an *HOXB6* gene variant, but abnormal expression patterns of the *HOXB6* gene have been observed in acute myeloid leukemia and colon cancer cells. And recently, an interesting study reported DNA hypermethylation in the *HOXB6* gene in the whole blood cell DNA samples from Alzheimer Disease (AD) patients.

6.4.2 PRECISION WITH REDUNDANCY

We see from these observations across the species spectrum from the fruit flies to mice and to humans the conservation and the adaptation of a "body plan" determination gene. The flies require the gene for embryonic survival but mice do not. The missing ribs in the *Hoxb6*-deficient mice mimicked the legs to antennae phenotype in the flies when the *Antp* gene was missing in the right place and at the right time. The *Hox* gene cluster in the flies has eight genes in a row, expressed in order by the segments from the "head" (labial) down to the "tail" (abdominal), in which the *Antp* comes in at the fifth position, governing the "mid-section" (central) where the second leg pair is located. A mutation that stops its properly placed and timed expression changes the second leg pair into antennae, whereas a mutation that induces its expression in the head changes the antennae into a leg pair.

The Hox genes in the mammal are described as "paralogous," that is, duplicated. The eight single genes in the fruit flies are duplicated into two, three, or four copies in four "banks": HoxA, HoxB, HoxC, and HoxD, with a total gene count of 39. The fly *Antp* is now *Hoxa6*, *Hoxb6*, and *Hoxc6*. The duplicated genes offer redundancy but also a more precise regulation of the "body plan." This redundancy can explain the partial phenotype in the *Hoxb6*-deficient mice which can develop to term and reproduce, but are imperfect with the skeletal defects. Mice with all three paralogous *Hox6* genes knocked out do not survive after birth. The mutant embryos do not appear or weigh differently from the normal littermate control mice, but have internal organ defects displaying an abnormal pancreatic morphology with the loss of insulin-secreting β cells.

6.4.3 CHANGES ARE NOT TOLERATED

What are the main lessons that we take away from the conserved and also diversified functions of the *Antp*, *Hoxb6*, and HOXB6 across the species from the fruit flies to mice and then humans? The first thing we can look at is the genetic text information, the DNA/gene sequences and the protein/amino acid sequences. In the 60-amino acid homeodomain, the namesake of the structural motif in the homeobox genes, the human HOXB6 protein and the mouse *Hoxb6* protein share 100% identity, while they also share 93.3% (56/60) identity with the *Drosophila Antp* protein. At the coding sequence level, the human and the mouse genes share 92.2% identity in the homeobox, and the identities between the human and the *Drosophila* genes, and between the mouse and the *Drosophila* genes are 81.1% and 79.4%, respectively.

To keep matters and our discussions simple and focus sharp, this high degree of conservation in the protein domain sequence can be interpreted as: changes are not tolerated. And even when the DNA sequence changed over the evolutionary course from the mouse to the human gene, with an ~8% difference in between, the amino acid sequence remained 100% conserved. The DNA sequence changes were on the third base of the codons and tolerated because they did not affect the encoded amino acids. It is remarkable that the protein sequence difference between the two mammalian species and the insect species is only less than 7%.

6.5 "What Will Modern Day Humans Evolve to?" and "Why Are Mutations Caused by Radiation or Chemicals Always Bad for Us?"

Can we use these results and interpretation from this sequence comparison exercise to answer the questions that we asked at the beginning of this chapter? Let's make an attempt.

The term "homeosis" was coined by William Bateson to theorize the underlying mechanistic basis of discontinuity in the origin of species, quite literally, "something has been changed into the likeness of something else." The change here is in reference of something to the effect of a body part, just like the legs to antennae and the antennae to legs in the fruit fly. And it literally only required the mistimed and misplaced expression of one single gene.

The change at the DNA/gene level in the expression regulation region of one single gene changed something at the physical or physiological level in an insect into the likeness of something else. This something else did not exist before the molecular change and is different from the original species both molecularly and physically. At this point the intent of Bateson's coining of the term homeosis can be considered fulfilled, that a pathway does seem to exist for creating new species in a discontinuous way. But the requirement for a new species to be reproductively insulated from the original species is not quite fulfilled.

6.5.1 GAIN OR LOSS OF FUNCTION

We dig a little deeper into the *Antp* gene mutations. Some are both dominant and recessive, with two different outcomes—dominant when in a heterozygote transforming the antennae into legs and recessive when in a homozygote causing lethality. Some only give the recessive lethality whereas some only cause the antennae transformation. As dramatic and form-changing as legs on the head without the antennae, it is clear that the flies carrying the *Antp* gene mutations are not a new species. They are the form variants within the limits of tolerance toward the genetic changes, including chromosomal rearrangements, as in the case of some *Antp* mutations.

An important point of consideration is the fact that most of these mutant fruit flies are kept in the laboratory vessels to remain viable and fertile for crosses in genetic experiments. In a natural environment or the wilderness, the loss of the sensory organ antennae can only be a huge disadvantage. These flies would not have much chance of surviving to develop and collect additional form-changing mutations to attain the status of a new species.

The genetic term "loss of function" is rather simple and straightforward. A mutation causes a gene not to be expressed or a variant form to be expressed without the normal function. If that function is required early in development, the result would be lethal in a homozygous state. In a diploid animal if the gene expression level from a single normal allele is sufficient, the effect would not be detected in a heterozygous state. This mutated allele would be considered a classic Mendelian recessive allele. If the gene expression level from a single normal allele is insufficient, that is, "haploinsufficiency," a heterozygote would also display a phenotype, somewhere between homozygote lethality and normal. And the trait would appear as classic Mendelian dominant. A loss of function takes away something from an organism after evolving and adapting over a long course of time and would result in perturbation of the perfect balance between the physiology of the organism and the environment. It is thus not surprising that many human developmental disorders are caused by haploinsufficient mutations in various transcription factor genes. If the haploinsufficiency involves reproduction, the new mutant allele would be finished off in the last carrier and not be able to go on passaging further.

The genetic term "gain of function" can be somewhat misleading with the use of the word "gain." The *Cambridge Dictionary* defines the word as "to get something that is useful, that gives you an advantage, or that is in some way positive, especially over a period of time." But in genetics and molecular biology, the "gain" of function loses the positive sense and refers simply and merely to "a molecule having a different function." We have mentioned the examples of the alcohol dehydrogenase *ADH1B*2* variant and the *ADH1B*3* variant that increase the alcohol to aldehyde turnover rate by 70–80 fold. The conversion of ethanol to ethylaldehyde is the rate-limiting step of alcohol metabolism, followed by the ethylaldehyde conversion to acetic acid. There is a conditional recognition of such an impressive gain of function in the alcohol dehydrogenase variants as beneficial mutations if the accumulated aldehyde does not overwhelm the enzymes of the next step, the aldehyde dehydrogenases. On the other hand, if the accumulation of aldehyde causes the adverse effects and limits the carriers from binge drinking, it is also beneficial but in an ironic way.

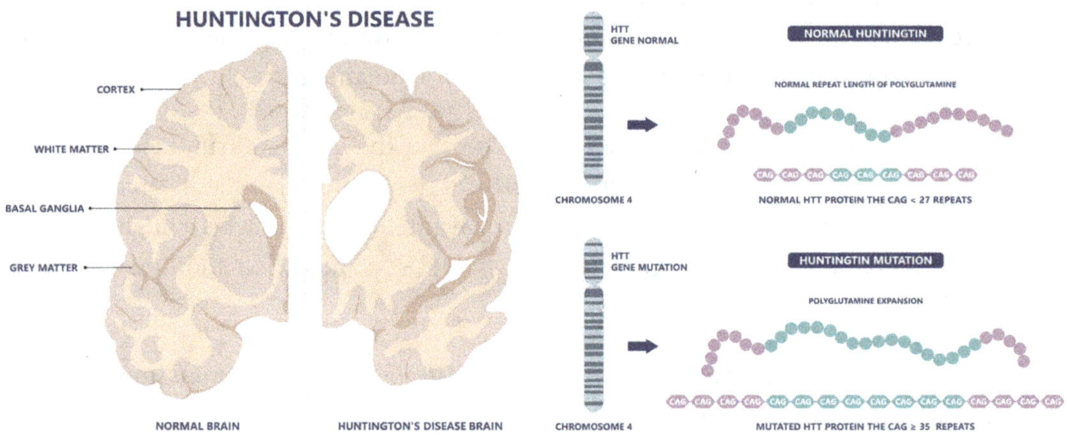

Pepermpron/Shutterstock.com

FIGURE 6-9. Huntington disease (HD) pathology and genetics

In human genetic disorders, Huntington Disease (HD) is the model for an autosomal dominant disease. HD is a progressive neurodegenerative disease, affecting motor, psychiatric, and cognitive functions and behaviors. The prevalence of HD in the general population is ~14 *per* 100,000 with no cure at the present time. The HD gene is named *HTT* (huntingtin), located on human chromosome 4p16.3, at a length of 180 kbp (kilo base pairs) with 67 exons. The HD allele is an expanded, unstable (CAG)n trinucleotide repeat at $n \geq 40$ in the first exon, which translates to a polyglutamine (poly-Q) repeat in the huntingtin protein (**Figure 6-9**). Longer repeats correspond to earlier disease onset by age. A repeat number $n < 36$ is considered normal or healthy, and repeat numbers between 36 to 39 show incomplete penetrance of HD, that is, some display the disease phenotype but some do not. The normal HTT protein is expressed throughout the body, including the brain. The HD HTT variant in the brain, however, displays an altered cellular localization than that of the normal protein's. The normal huntingtin protein acts as an intracellular scaffolding protein, with functions in cellular protein transport. More than 700 potential interactive protein partner candidates have been identified. Currently HD is managed clinically with drugs that alleviate the symptoms but no major advancements toward a cure have been made in the past 20 years. New molecular therapies are being developed to target the variant HTT mRNA or the variant allele DNA. But the road ahead presents more obstacles than offering hope and promise.

The mouse *Htt* gene shares 86% sequence identity with the human *HTT* gene. The mouse Htt protein shares 91% identity at the translated protein sequence level with the human counterpart. Curiously, the mouse *Htt* (CAG)n repeat is only 7 units long (with the third repeat being CAA), in contrast to the 13 to 36 units in the normal human alleles. There is no evidence for length variations of the (CAG)7 repeat in mice, and there is no spontaneous mouse model of HD. An insertion of a 146-unit CAG repeat into the mouse hypoxanthine phosphoribosyltransferase gene (*Hprt*) lead to a late-onset neurological disease with premature death, whereas inactivation of the *Hprt* gene itself did not have any deleterious effects in mice. Homozygote knockout of the mouse *Htt* gene is embryonic lethal whereas heterozygote *Htt* knockout is associated with neurological and behavioral phenotypes. Targeted *Htt* gene deletion in a tissue-specific manner in the forebrain and testis of adult mice results in progressive neurodegeneration, sterility, and reduced lifespan.

It would be naïve to strictly describe HD as a gain-of-function dominant disease, as the functional gain of the disease variant protein is not exactly clear at the molecular mechanistic level. It has been known as toxic because of its altered cellular distribution and its ability to recruit the normal protein into insoluble aggregates, very much like a "dominant negative." This term describes a protein variant with a loss of the normal function but "gaining" the property and ability to interfere with the proper function of the normal protein in the context of a heterozygote. The gene targeting experiments in mice also provide support that the loss of one *Htt* allele produces phenotypes similar to HD.

6.5.2 DELETERIOUS MUTATIONS THAT GAINED MOLECULAR FUNCTIONS

Although beneficial "gain-of-function" mutations are difficult to list, deleterious examples are many. The epidermal growth factor receptor (EGFR) is a transmembrane receptor tyrosine kinase (RTK), with the growth factor receptor domain on the outside and the tyrosine kinase domain (TKD) on the inside of the cell (**Figure 5-3**). The receptor is a homodimer, that is, formed by two identical units, and exists in a conformation that mutually inhibits the two kinase domains inside the cell. Engagement of the epidermal growth factor (EGF) changes the conformation and activates the TKD on both monomer EGFRs, which reciprocally phosphorylate an array of tyrosine residues. The phosphorylated tyrosine residues serve as docking sites for downstream signaling molecules to initiate a signal transduction cascade, culminating with receptor signal-specific gene expression and cell cycle activation. This signaling axis is tightly regulated by receptor inhibition in the absence of the growth factor and turnover of the growth factor-engaged receptor, called ligand-induced degradation. Only the cells that receive the proper growth factor signal, for example, during wound repair after tissue damage, are allowed to have a transient presence of activated EGFR.

Dysregulation of the EGFR signaling is a major cause of lung cancer (non-small cell lung cancer, NSCLC). The receptor molecule can be overexpressed, and the overabundance can overcome the conformational tyrosine kinase inhibition in the receptor dimer. Many mutations have been reported for the EGFR with the phenotype of "constitutive activation," that is, activation in the absence of the growth factor. Such constitutively activated EGFR variants drive cell proliferation, angiogenesis, invasion, and metastasis. Most are somatic mutations, including deletions, insertions, and substitutions. Some are germline, for example, the T790M variant, first identified in a family of European descent with multiple members developing lung adenocarcinoma. The constitutively active EGFR variants are the molecular targets of therapy for lung cancer using small molecule inhibitors known as tyrosine kinase inhibitors (TKIs), such as gefitinib (Iressa) and erlotinib (Tarceva).

6.5.3 HUMAN EVOLUTION IN THE FUTURE

Returning to Bateson's vision with the coining of the term homeosis, theorizing the possibility of discontinuous speciation by changing something to the likeness of something else. The problem is, for the organisms that are in the current forms of perfect harmony and balance between their physiology and their environments, losses or gains of functions at the molecular level do not appear to have realistic possibilities of bringing about benefits. One current opinion on the "must" of human evolution argues for it from the reverse angle, that the ideal population of infinitely numerous members with no mutation, no selection, and no migration and with completely random mating does not and cannot exist. Using the definition of "allele frequency change" or "new alleles" in the genome as a definition of evolution, the answer is an emphatic yes.

Using phenotypes? Some examples given include that the average height of humans has considerably increased over the past 150 years and that the short-sightedness has also significantly increased in the civilized societies. These phenotypes are, however, mostly a reflection of the environmental and cultural changes in the recent times, and still within the limits of tolerance toward the genetic or epigenetic changes. On the other hand, future space travel with colonizing human populations on a different planet appears to be an untested but the most convincing way to envision human evolution to the extent of diverging to new species.

BIBLIOGRAPHY

1. https://www.genome.gov/about-genomics/fact-sheets/Cloning-Fact-Sheet
2. https://www.omim.org/entry/613004#title
3. Anvar Z, Chakchouk I, Demond H, Sharif M, Kelsey G, Van den Veyver IB. DNA methylation dynamics in the female germline and maternal-effect mutations that disrupt genomic imprinting. *Genes* (Basel). 2021 Aug 6;12(8):1214.
4. Armstrong A, Kroener L, Miller J, Nguyen A, Kwan L, Quinn M. The nature of embryonic mosaicism across female age spectrum: An analysis of 21,345 preimplantation genetic testing for aneuploidy cycles. *F&S Reports*. 2023 Apr 8;4(3):256–61.
5. Ashok A, Pooranawattanakul S, Tai WL, Cho KS, Utheim TP, Cestari DM, Chen DF. Epigenetic regulation of optic nerve development, protection, and repair. *International Journal of Molecular Sciences*. 2022 Aug 10;23(16):8927.
6. Barron JC, Hurley EP, Parsons MP. Huntingtin and the synapse. *Frontiers in Cell Neuroscience*. 2021 Jun 15;15:689332.
7. Blackman VH. John Bretland Farmer. 1865–1944. Obituary notices of Fellows of the Royal Society. 1945;5(14):17–31. http://www.jstor.org/stable/769107
8. Cagan A, Baez-Ortega A, Brzozowska N, Abascal F, et al. Somatic mutation rates scale with lifespan across mammals. *Nature*. 2022 Apr;604(7906):517–524.
9. Cheng H, Shang D, Zhou R. Germline stem cells in human. *Signal Transduction and Targeted Therapy*. 2022 Oct 2;7(1):345. doi:10.1038/s41392-022-01197-3. Erratum in: *Signal Transduction and Targeted Therapy*. 2022 Dec 3;7(1):385.
10. Condic ML. Totipotency: What it is and what it is not. *Stem Cells Development*. 2014 Apr 15;23(8):796–812.
11. Coticchio G, Cimadomo D, Cermisoni GC, Rienzi L, Papaleo E, Ubaldi FM, Borini A, De Santis L. The first mitotic division: A perilous bridge connecting the zygote and the early embryo. *Human Reproduction*. 2023 Jun 1;38(6):1019–27.
12. Currie CE, Ford E, Benham Whyte L, et al. The first mitotic division of human embryos is highly error prone. *Nature Communications*. 2022;13:6755.
13. Ferguson MW, Kennedy CJ, Palpagama TH, Waldvogel HJ, Faull RLM, Kwakowsky A. Current and possible future therapeutic options for Huntington's disease. *Journal of Central Nervous System Disease*. 2022 May 21;14:11795735221092517.
14. Ferguson-Smith AC, Bourc'his D. The discovery and importance of genomic imprinting. *Elife*. 2018 Oct 22;7:e42368.
15. Johnson AD, Alberio R. Primordial germ cells: The first cell lineage or the last cells standing? *Development*. 2015 Aug 15;142(16):2730–9.

16. Kanherkar RR, Bhatia-Dey N, Csoka AB. Epigenetics across the human lifespan. *Frontiers in Cell and Developmental Biology*. 2014 Sep 9;2:49.
17. Kappen C. Disruption of the homeobox gene Hoxb-6 in mice results in increased numbers of early erythrocyte progenitors. *American Journal of Hematology*. 2000 Oct;65(2):111–8.
18. Kermi C, Aze A, Maiorano D. Preserving genome integrity during the early embryonic DNA replication cycles. *Genes* (Basel). 2019 May 24;10(5):398.
19. Khan YS, Ackerman KM. Embryology, Week 1. [Updated 2023 Apr 17]. In: *StatPearls* [Internet]. Treasure Island, FL: StatPearls Publishing; 2024 Jan. Available from: https://www.ncbi.nlm.nih.gov/books/NBK554562/
20. Kun Á. Is there still evolution in the human population? *Biologia Futura*. 2022;73:359–74.
21. Larsen BM, Hrycaj SM, Newman M, Li Y, Wellik DM. Mesenchymal Hox6 function is required for mouse pancreatic endocrine cell differentiation. *Development*. 2015 Nov 15;142(22):3859–68.
22. Lebecque B, Bourgne C, Vidal V, Berger MG. DNA methylation and intra-clonal heterogeneity: The chronic myeloid leukemia model. *Cancers* (Basel). 2021 Jul 17;13(14):3587.
23. Lewis EB. Homeosis: The first 100 years. *Trend in Genetics*. 1994 Oct;10(10):341–3.
24. Li D, Wintle NJP, Zhang G, Wang C, Luo B, Martin-Wintle MS, Owen MA, Swaisgood RR. Analyzing the past to understand the future: Natural mating yields better reproductive rates than artificial insemination in the giant panda. *Biological Conservation*. 2017;216:10–17.
25. Lin M, Sigal M. Human embryo models: Unveiling sophisticated self-organization of stem cells during post-implantation stages. *Signal Transduction and Targeted Therapy*. 2023 Nov 6;8(1):419.
26. Lowe J, Curry E. Incidence of pregnancy loss and characterization of fetal development in red pandas. *Reproduction and Fertility*. 2021 Oct 21;2(4):292–9.
27. Masset H, Tšuiko O, Vermeesch JR. Genome-wide abnormalities in embryos: Origins and clinical consequences. *Prenatal Diagnosis*. 2021 Apr;41(5):554–63.
28. Myers, P. Hox genes in development: The Hox code. *Nature Education*. 2008;1(1):2. http://scienceblogs.com/pharyngula/2007/09/the_hox_code.php
29. Oxnard GR, Chen R, Pharr JC, Koeller DR, et al. Germline EGFR Mutations and familial lung cancer. *Journal of Clinical Oncology*. 2023 Dec 1;41(34):5274–84.
30. Paine H. Does loss of the normal protein function contribute to the pathogenesis of Huntington's disease? *Bioscience Horizons*. 2015;8.
31. Paro R, Grossniklaus U, Santoro R, et al. Introduction to epigenetics [Internet]. Cham, CH: Springer; 2021. Chapter 5, Genomic Imprinting. 2021 Mar 24. Available from: https://www.ncbi.nlm.nih.gov/books/NBK585708/
32. Paulson JR, Hudson DF, Cisneros-Soberanis F, Earnshaw WC. Mitotic chromosomes. *Seminars in Cell & Devopmental Biology*. 2021 Sep;117:7–29.

33. Rossant J, Tam PPL. Early human embryonic development: Blastocyst formation to gastrulation. *Developmental Cell.* 2022 Jan 24;57(2):152–65.
34. Shahbazi MN. Mechanisms of human embryo development: From cell fate to tissue shape and back. *Development.* 2020 Jul 17;147(14):dev190629.
35. Schneuwly S, Gehring WJ. Homeotic transformation of thorax into head: Developmental analysis of a new Antennapedia allele in *Drosophila melanogaster. Developmental Biology.* 1985;108(2):377–86.
36. Schneuwly S, Klemenz R, Gehring WJ. Redesigning the body plan of Drosophila by ectopic expression of the homoeotic gene Antennapedia. *Nature.* 1987 Feb 26–Mar 4;325(6107):816–18.
37. Schorsch M, Gomez R, Hahn T, Hoelscher-Obermaier J, Seufert R, Skala C. Success rate of inseminations dependent on maternal age? An analysis of 4246 insemination cycles. *Geburtshilfe Frauenheilkd.* 2013 Aug;73(8):808–11.
38. Singh A, Rappolee DA, Ruden DM. Epigenetic reprogramming in mice and humans: From fertilization to primordial germ cell development. *Cells.* 2023 Jul 17;12(14):1874.
39. Zug R. Developmental disorders caused by haploinsufficiency of transcriptional regulators: A perspective based on cell fate determination. *Biology Open.* 2022 Jan 15;11(1):bio058896.

7

Finding "Disease Genes" by Genetic Linkage Analysis

The comedian Jeff Foxworthy told this joke in his Totally Committed act in 1998: "And talking about losing things, I loved this one . . . 'it was in the last place I looked.'"

In the search of a disease gene, still a misnomer and should have been "a gene with a variant allele that causes a disease" (and that's why we simply say a "disease gene"), what is "the last place we look?"

The very last place we look is the locus where an allele has a 100% association with the disease phenotype and the other allele(s) have a 100% association with the normal phenotype. We have such examples in the sickle cell disease (SCD), for which an A to T mutation in the seventh codon of the β-globin gene changes a Glu to a Val and causes the disease in the T/T homozygote. In the Huntington Disease (HD), an expansion of the (CAG)n trinucleotide repeat in the first exon of the

HTT gene to greater than 40 units ($n \geq 40$) is a diagnostic marker for disease onset. In cystic fibrosis (CF), a deletion of the triplet codon for Phe/F at position 508 (ΔF508) is causative for the disease in a ΔF508/ΔF508 homozygote. The route to the last place for finding the *SCD* gene started with the analysis of the encoded protein, β-globin, and ended with the cloning and localization of the β-globin gene on human chromosome 11. We described this as "forward genetics," as in marching forward to a genetic resolution of the disease.

For HD or CF, the disease was very well characterized clinically but there was no information or method to speculate on the causative variant protein, assuming that a variant protein was the cause. In this regard, the way forward would be not at all different from what Gregor Mendel or Thomas Morgan and his student colleagues were doing, working with expressed traits with no information on the genes. Before the completion of the Human Genome Project and the work under the 1000 Genomes Project, genetic markers that only differ at the DNA level but without physiological manifestations were not readily available. As late as the early 1990s, for the particular example of the very first human genetic disorder to be described, alkaptonuria, the mapping of its gene (*HGD*) relied on the location of another disease gene (*FHH*) mapped prior. That was quite literally genetics practiced in the fashion of Mendel or Morgan.

7.1 The Number of Progeny Required for Classical Genetic Studies

7.1.1 RECOMBINATION FREQUENCY/FRACTION INFORMS ON LOCUS–LOCUS RELATIONSHIP IN GENETIC STUDIES

The law of genetics was first investigated by and credited back to Mendel as "independent segregation," describing the true relationship of alleles at a locus and at multiple loci. The lack of independent segregation of alleles across two loci observed in the fly lab of Morgan's offered the only measurement of locus–locus relationship by the number of trait recombinants at the two loci. From what we have reviewed in the previous chapters, when the recombination fraction (RF) is 0.5 (recombination frequency at 50%), no conclusion can be drawn on whether the two loci are located on the same chromosome or on two different ones. When RF is <0.5 (recombination frequency < 50%), the two loci are on the same chromosome. To determine the RF, a good number of pea seeds or fruit flies are required to give the rigor in the statistical tests on whether the recombinants are at 50% or less. Mendel relied on less than 100 plants and Morgan and his colleagues routinely counted about 1,000 flies. As the "genetic distances" of two loci on a chromosome are correlated to the numbers of recombinants, the numbers in the fly experiments were good enough to offer a resolution of 5% while maintaining the statistical rigor. At the physical chromosomal DNA distance level, that is about 5 million base pairs.

The world record of the same couple having the largest number of children, 69, belonged to the Russian couple Feodor Vassilyev and Valentina Vassilyeva. The 69 came from 16 pairs of twins, 7 sets of triplets, and 4 sets of quadruplets, so the actual number of meiosis being represented was 27, still very impressive. Today, the average number of children *per* family is around two. These numbers do not compare to the requisite numbers in the fruit flies for statistical rigor in any attempt of determining the RF, in truth, the departure of RF from 0.5.

7.1.2 THE PROBABILITIES OF INDEPENDENT MEIOTIC EVENTS

In **Chapter 2** we have compared this to the exercises of coin toss. Let's propose that the goal of the coin tosses is to determine if the coin is even-sided, which would give a 50% *versus* 50% heads to tails distribution. But there are two conditions. First, the tossing has to be done consistently without bias, such as by a mechanical device. And second, still, there needs to be a high enough number of tosses. For example, tossing twice and getting the same side up cannot be used to accuse the coin of being uneven. Intuitively, tossing a coin 100 times and getting 49:51 would be good enough to prove that it is even. And getting 30:70 would reveal a problem of the mold in which the coin was minted. Tossing one coin 100 times by the exact same method is directly comparable to examining the allele–allele relationships across two loci in 100 children of a single family.

We now use 10 coins from the same mold on 10 identical mechanical devices to toss 10 times on each device. For strictly the coin toss exercise, we can lump all the heads and tails together and make a determination. This situation would be the same as examining the allele–allele relationships across two loci in 10 children *per* single family, over 10 families. We can think of the Reference Human Genome as the mold, and each parent in the families as the carrier of the alleles printed off the Reference Human Genome, the coins. The alleles across the two loci being examined either recombine or stay in the same linkage phase during meiosis, manifested and reported by the offspring. Intuitively, we should see no reason for us not to be able to lump the recombinants and the nonrecombinants from the 100 children in the 10 different families.

7.1.3 HUMAN FAMILY VARIATIONS IN THE GENETIC OUTPUT

But there is a reason. Unlike the divided coin toss being performed on 10 identical mechanical devices, the human reproductive physiology is not identical from individual to individual, or from population to population, and the rates of recombination can vary due to differences in the intrinsic, as well as environmental factors. Not all regions of the genomic DNA have the same propensity of recombination during meiosis. Rather, the crossing over of the nonsister chromatids occur at "hot spots." The protein PRDM9 (PR domain containing protein 9) is a DNA-binding, zinc finger, histone methyltransferase which determines the location of the recombination hotspots in both male and female germ cells in humans and mice. PRDM9 is called "a driver of the genetic map," and variations in the *PRDM9* gene causes variations in the locations and activities of the recombination hotspots across individuals and populations, and between male and female. This is one but an impactful example to underscore the lack of a consistent tosser of the alleles across any two loci and the reason why recombinants and nonrecombinants from different families cannot be lumped to calculate the RF.

7.2 The LOD Score Analysis

The dilemma we face is that the only way for us to determine the relationship between the two loci, an unknown disease locus and a known marker locus, is to calculate the RF based on the number of recombinant *versus* nonrecombinant offspring. From a single family, the numbers are not enough. From multiple families, because of the various intrinsic and extrinsic factors that affect the molecular mechanisms in the meiotic chromosome crossover, the numbers cannot be lumped. This is like having 10 identical coins being tosses by 10 different persons, with each person using a different technique of tossing.

When we look at a family of 10 children ($n = 10$), presenting four recombinants ($k = 4$) and six nonrecombinants ($n - k = 6$) for the disease locus D/d and a marker locus M1/M2, there are two possible relationships between the two loci. One, they are genetically linked and the RF (θ) is <0.5. Two, they are not genetically linked, that is, being on two different chromosomes or far apart on the same chromosome, and the RF (θ) is 0.5. For each child presenting the allele recombination across the two loci, the probability is θ. To have k recombinant children, the probability is θ^k. Likewise, to have $(n - k)$ nonrecombinant children, the probability is $(1 - \theta)^{n-k}$. The probability of getting the specific genotypes in all the children is $\theta^k(1 - \theta)^{n-k}$.

For the first possibility that the two loci are genetically linked, θ provides the answer of whether this is true, but it is unknown. For the second possibility that the two loci are not genetically linked, the probability of getting the specific genotypes in all the children is 0.5^n. The question now becomes, which possibility is more likely or how do we tell which possibility is more likely? We calculate the odds, and for cleanliness of data presentation, we look at the $\log_{10}$ (lg) of the odds, known as a LOD score:

$$LOD = \lg \left[\frac{\theta^k(1-\theta)^{n-k}}{0.5^n}\right]$$

All is good but the problem is, we do not know what to use as the value of θ. For the hypothetical family of 10, with four children presenting the recombinant genotypes, is it 4/10 = 40% or RF (θ) = 0.4? We cannot state that. If we could, none of the LOD score discussion is necessary. But we know something, that RF (θ) is between 0 and 0.5. Here we use the coin toss comparison one last time: if RF (θ) is 0.5, the coin is even, and if RF (θ) is 0, the coin lands on the same side every time. In the data from a single person's toss, we cannot distinguish between a bad coin and a bad tosser. Likewise, from a single family's multiple independent telltale meiosis events, that is, gametes that presented as the genotypes in the children, we cannot distinguish between linked recombination and independent assortment.

7.2.1 LOD SCORES FROM FAMILIES OF 10 CHILDREN EACH WITH WIDESPREAD k VALUES

We can calculate the LOD scores over the full range of RF (θ) values and examine the data to see what they will inform us. **Figure 7-1** shows the data from four different families of 10 children each with different numbers (*k*) of children presenting the recombinant genotypes.

Let us look at and interpret the data. First, all curves end at LOD = 0 when RF (θ) = 0.5, which is the state that the top and the bottom of the LOD score equation are the same, and lg 1 = 0. On the other side, when RF (θ) = 0.0, the genetic definition is: no recombinants are observed. But the *k* value is not zero from any family, and the hypothesis directly contradicts the facts. Thus the LOD scores are negative infinity for all families. The interpretation is: the odds for not observing recombinants are impossible.

The blue curve ($k = 2$) represents the family with two recombinants, and the apparent RF (θ) would be 0.2. At RF (θ) = 0.2, the LOD score is 0.84, the highest point over the full range of RF (θ) values. This translates to "it is ~6.9 times more likely that the genotype presentations in the children were the results of linked recombination at a frequency of 20%," hardly a good odds to bet the mortgage on.

The red curve is the summary curve of all four families using the sum of all LOD scores at each RF (θ) value. The odds from multiple independent events are multiplicable, thus the LOD scores are additive. We cannot extract much information from the red curve, but we can state the following.

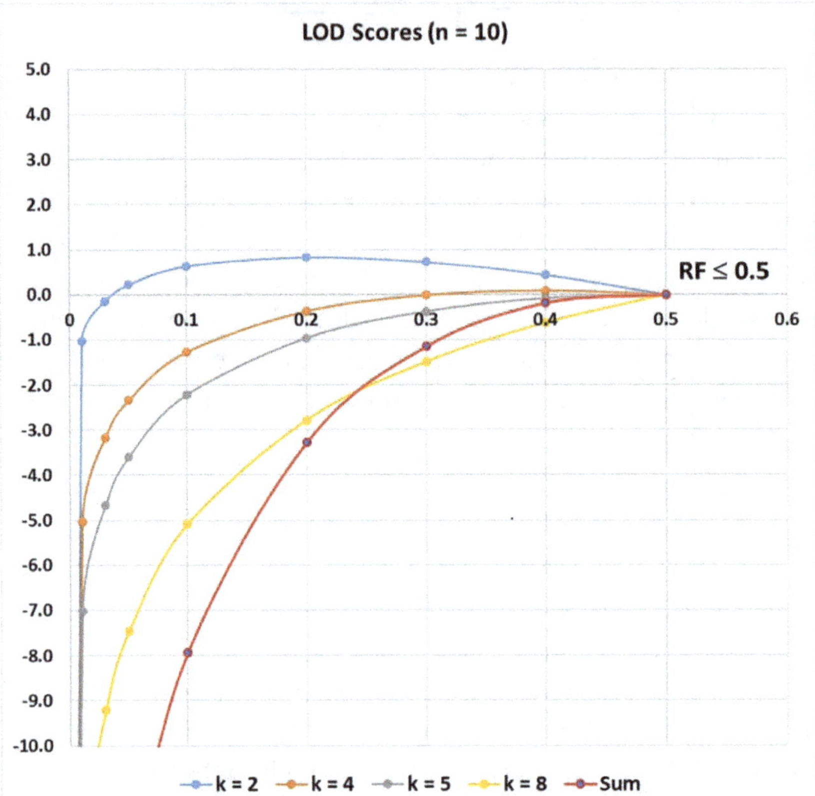

Model n = 10	0	0.01	0.03	0.05	0.1	0.2	0.3	0.4	0.5
k = 2	-∞	-1.02	-0.14	0.23	0.64	0.84	0.73	0.44	0
k = 4	-∞	-5.02	-3.16	-2.33	-1.26	-0.37	-0.01	0.09	0
k = 5	-∞	-7.01	-4.67	-3.6	-2.22	-0.97	-0.38	-0.09	0
k = 8	-∞	-13	-9.2	-7.44	-5.08	-2.78	-1.48	-0.62	0
Sum	-∞	-2.61E+01	-1.72E+01	-1.31E+01	-7.92E+00	-3.28E+00	-1.14E+00	-1.80E-01	0.00E+00

FIGURE 7-1. LOD scores from families of 10 children each with widespread *k* values

Let's break our own rule and lump the recombinants together, so in all 40 children there are 19 recombinants, nearly half. The apparent RF (θ) is 19/40 = 0.475. So at every RF (θ) value the LOD score is negative, the smaller the RF (θ) value, the more negative the LOD score. And the odds are even at RF (θ) = 0.5. The genetic linkage study on the two loci ends with an ambiguous answer that they could be on two different chromosomes or far apart on the same chromosome. We could have gotten this ambiguous answer by the 19:21 recombinant *versus* nonrecombinant ratio without calculating all the LOD scores.

7.2.2 LOD SCORES FROM FAMILIES OF 10 CHILDREN EACH WITH SMALL k VALUES

Now let's look at four families with two children in each presenting the recombinant genotypes. The curves are identical so they collapse into the yellow curve (**Figure 7-2**). The summary red curve now peaks at RF (θ) = 0.2 with a LOD score of 3.36, that is, 2,291 times more likely that the two loci are genetically linked with a recombination frequency of 20%. Interestingly, this recombination frequency is exactly the same as the result if we lumped all the family data together: 8/40 = 20%. We did not ask a question on who the families were, where they were, or when in the history of time their mating outcomes with the genotype information were recorded. With an odds of 2,291 to 1, it looks like a good bet for the two loci to be genetically linked.

7.2.3 LOD SCORE THRESHOLDS FOR LINKAGE AND EXCLUSION OF LINKAGE

The null hypothesis being tested is that the two loci are not genetically linked with an RF (θ) = 0.5. For the linkage analysis using the LOD score as described above, known as parametric linkage analysis based on an established mode of inheritance, a LOD score of 3 or odds of 1,000:1 is equivalent to the commonly accepted p value of 0.05 for rejecting the null hypothesis. This LOD score threshold was set after considerations and calculations on the random chance that two loci would colocalize on the same chromosome. A LOD score of <−2 is used as the threshold to exclude genetic linkage between the two loci.

Both the +3 and the −2 thresholds for rejecting nonlinkage and linkage are dependent on the specific RF (θ) value. Thus the interpretation over the full range of RF (θ) values may be different for the two loci. The interpretation on the plus side is rather straightforward and intuitive. For the minus side, in the summary curve of the first four families of 10 children each (**Figure 7-1**), the exclusion of linkage is strong when RF (θ) is <0.25, with LOD scores <−2. When the RF (θ) is at 0.3, the evidence is weak with the LOD score at −1.14. The evidence continues to weaken between RF (θ) values 0.3 to 0.5. This makes sense as the actual lumped recombinants were nearly half of the total (19/40), arguing strongly against linkage at recombination frequencies of 10%, 20%, and 25%. Between >25% to <50%, the genotype presentations in the four families may be different from the null hypothesis of 50% by chance but the available data do not allow a conclusion. In the summary curve of the second four families of 10 children each (**Figure 7-2**), the exclusion of linkage at RF (θ) <0.03 is strong, with a LOD score of −5.6. An RF (θ) = 0.03 translates to observing three recombinants in 100, but the four families already presented eight in total. A more literal translation would be, even if the two loci were genetically linked on the same chromosome, they are not that close to each other.

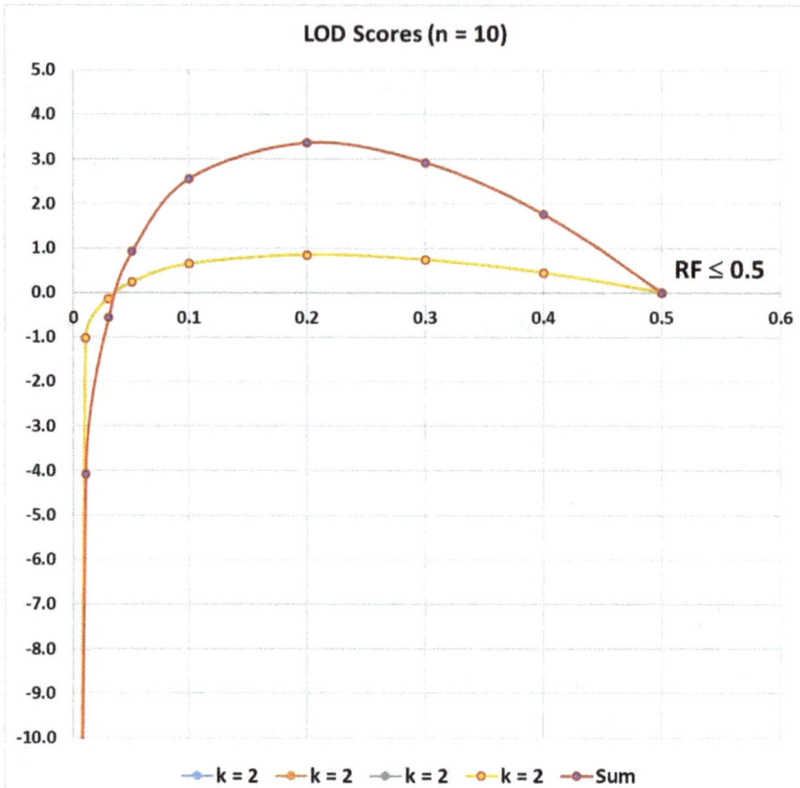

Model n = 10	0	0.01	0.03	0.05	0.1	0.2	0.3	0.4	0.5
k = 2	-∞	-1.02	-0.14	0.23	0.64	0.84	0.73	0.44	0
k = 2	-∞	-1.02	-0.14	0.23	0.64	0.84	0.73	0.44	0
k = 2	-∞	-1.02	-0.14	0.23	0.64	0.84	0.73	0.44	0
k = 2	-∞	-1.02	-0.14	0.23	0.64	0.84	0.73	0.44	0
Sum	-∞	-4.08E+00	-5.60E-01	9.20E-01	2.56E+00	3.36E+00	2.92E+00	1.76E+00	0.00E+00

FIGURE 7-2. LOD scores from families of 10 children each with small *k* values

7.2.4 LOD SCORES FROM THE FAMILY OF FEODOR AND VALENTINA

We now perform a hypothetical linkage analysis of two loci, a disease D/d and a genetic marker M1/M2, in the family of Feodor and Valentina. We calculate the LOD scores across the full range of RF (θ) values using k = 3, 7, 13; and n = 27 (number of independent meiosis on Feodor's side).

At $n = 27$, $k = 3$ (blue curve), the Vassilyev family produced a very strong evidence of linkage of the two loci at an RF (θ) = 0.1, with a LOD score of 4.03 (odds of 10,715:1), the highest point over the full range (**Figure 7-3**). Interestingly, this RF (θ) value determined by the LOD score linkage analysis is not too different from the actual value: 3/27 = 1/9 = 0.11. With $k = 7$ (orange curve), approximately 25%, the power of the LOD score test is not strong for linkage at any RF (θ) value but very strong for rejecting RF (θ) less than 0.04 (4%, far below the actual ~25%). At $k = 13$ (gray curve), approximately 50%, the curve looks very similar to the summary curve of the four families of 10 children each with widespread k values (**Figure 7-1**).

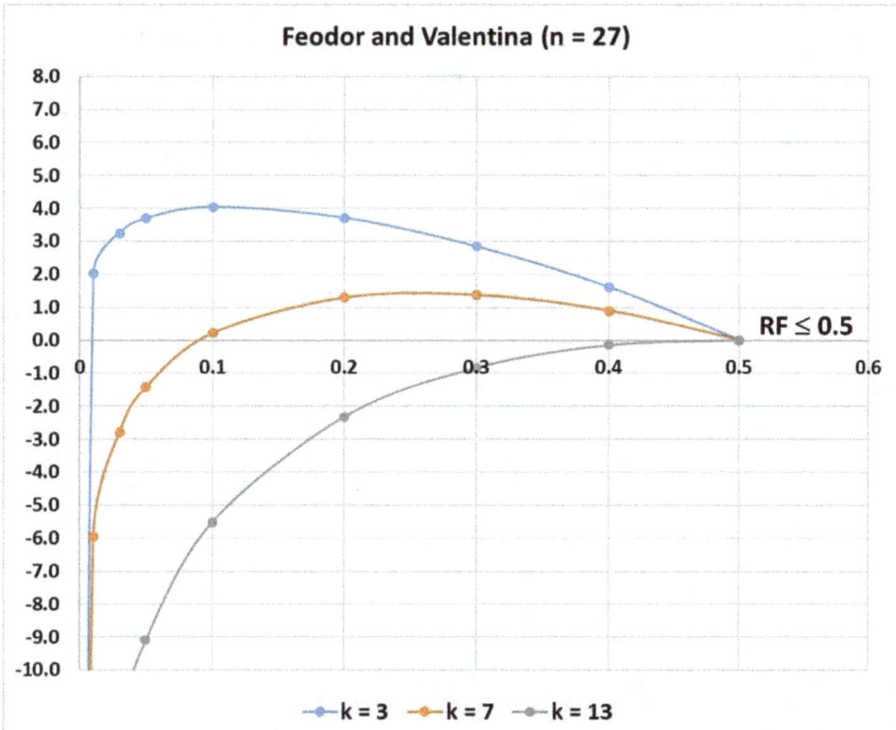

Feodor n = 27	0	0.01	0.03	0.05	0.1	0.2	0.3	0.4	0.5
k = 3	-∞	2.02	3.24	3.69	4.03	3.7	2.84	1.61	0
k = 7	-∞	-5.96	-2.8	-1.42	0.21	1.29	1.37	0.9	0
k = 13	-∞	-17.93	-11.85	-9.1	-5.51	-2.32	-0.84	-0.15	0

FIGURE 7-3. LOD scores from the family of Feodor and Valentina

7.2.5 LOD SCORES FROM MODERN NUCLEAR FAMILIES WITH ONE CHILD

Now we move to the other extreme in human families, the modern nuclear type with one child. Intuitively we would dismiss such families entirely for enrollment in a genetic linkage study to map the location of a disease gene using genetic markers. After all, in the first example of families with 10 children and widespread *k* values, the curves did not inform us much, except to reject the association of alleles across the two loci at RF (θ) < 0.25. We have drawn three summary curves for 10 families of one child each (**Figure 7-4**). In the first 10, five of the children from five different families were recombinant, in the second 10, all recombinant, and in the third 10, all nonrecombinant. The first two curves (blue and orange) do not inform us much, except to reject the association of alleles across the two loci at RF (θ) <0.1 for the blue curve, and <0.3 for the orange, due to the presence of the recombinants. The third curve, representing 10 children from 10 different families with zero recombinant, gives a strong threshold-clinching LOD score of 3 for RF (θ) = 0. The LOD scores are decreasing with the increasing θ values, which expect more and more recombinants, yet none was observed in 10 different families.

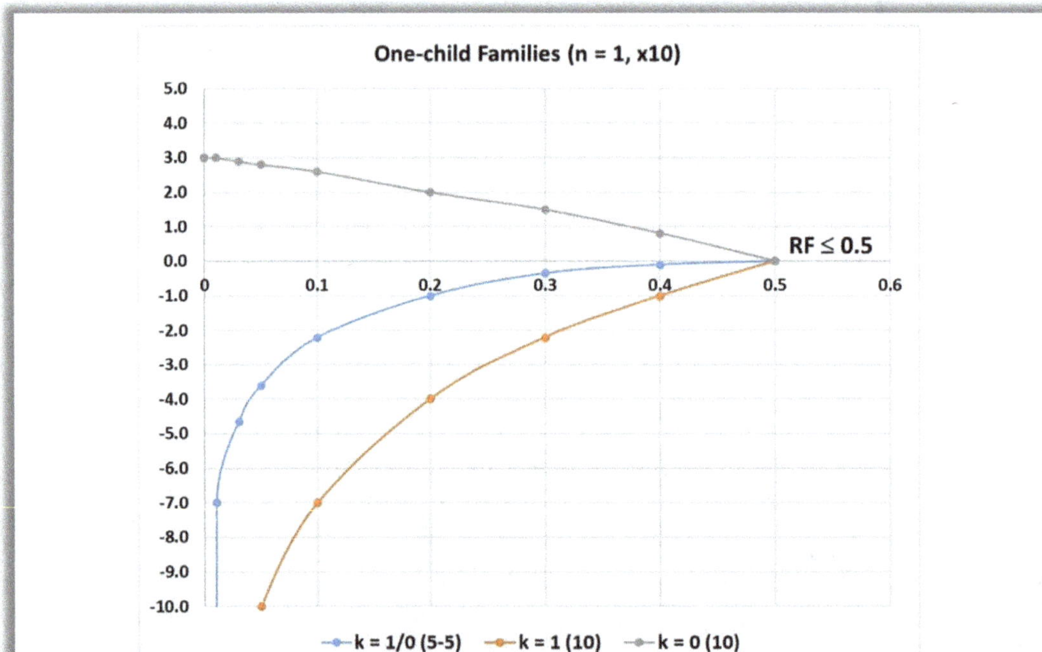

Model n = 1 (10)	0	0.01	0.03	0.05	0.1	0.2	0.3	0.4	0.5
k = 1/0 (5/5)	-∞	-7	-4.65	-3.6	-2.2	-1	-0.35	-0.1	0
k = 1 (10)	-∞	-17	-12.2	-10	-7	-4	-2.2	-1	0
k = 0 (10)	3	3	2.9	2.8	2.6	2	1.5	0.8	0

FIGURE 7-4. LOD scores from modern nuclear families with one child

7.2.6 THE INFORMATION ON LINKAGE PHASE HAS AN IMPACT ON LOD SCORES

In summary, we can see the following key points for a classic parametric linkage analysis using the LOD scores to map the chromosomal location of a disease gene relative to a genetic marker locus.

- The mode of inheritance of the disease should be known. Our preferred parents are a double-heterozygote father for the two loci D/d and M1/M2, and a double-recessive homozygote mother (the molecular marker locus genotypes are codominant in effect). Still preferably the linkage phase of the father is known, for example, he inherited D-M1 and d-M2 together from his two parents, to allow for the determination of recombinants. For this reason, multigeneration and large families with multiple children are preferred. The term "linkage phase" refers only to the coinheritance of alleles from a parent, but does not imply genetic linkage of the two loci.

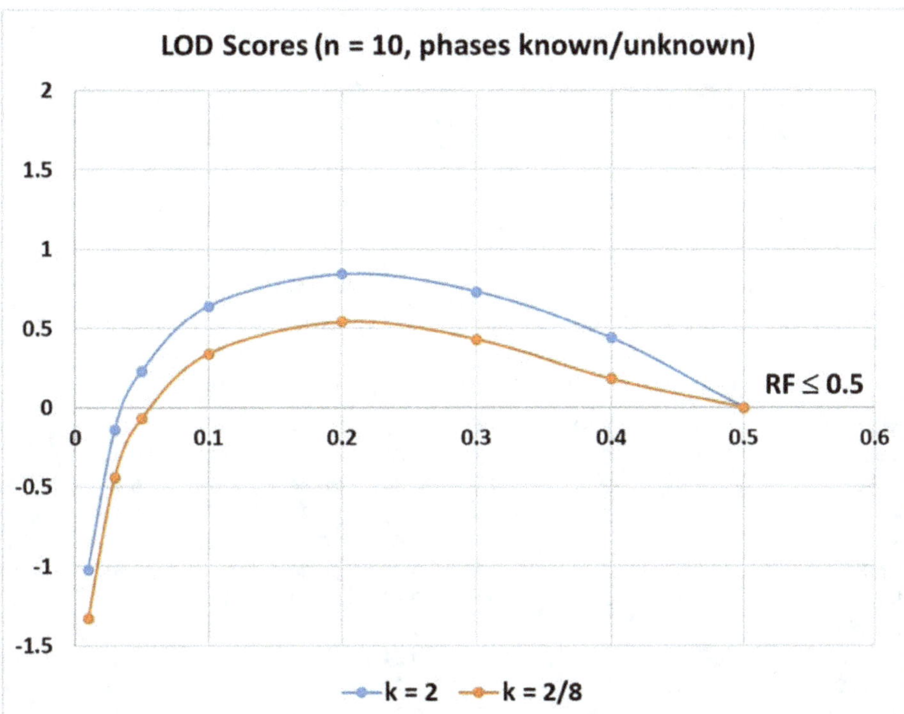

Model n = 10	0.01	0.03	0.05	0.1	0.2	0.3	0.4	0.5
k = 2	-1.02	-0.14	0.23	0.64	0.84	0.73	0.44	0
k = 2/8	-1.33	-0.44	-0.07	0.34	0.54	0.43	0.18	0

FIGURE 7-5. LOD scores from families with known *versus* unknown linkage phase

- Not knowing the linkage phase of the father prevents the precise determination of recombinants, but LOD scores can still be calculated. Essentially, the k and $n - k$ numbers are in two mutually exclusive conditions, with k representing the recombinants in one linkage phase and the nonrecombinants in the other linkage phase. The LOD score equation is revised to:

$$LOD = \lg \left[\frac{\theta^k(1-\theta)^{n-k} + \theta^{n-k}(1-\theta)^k}{0.5^{n-1}}\right]$$

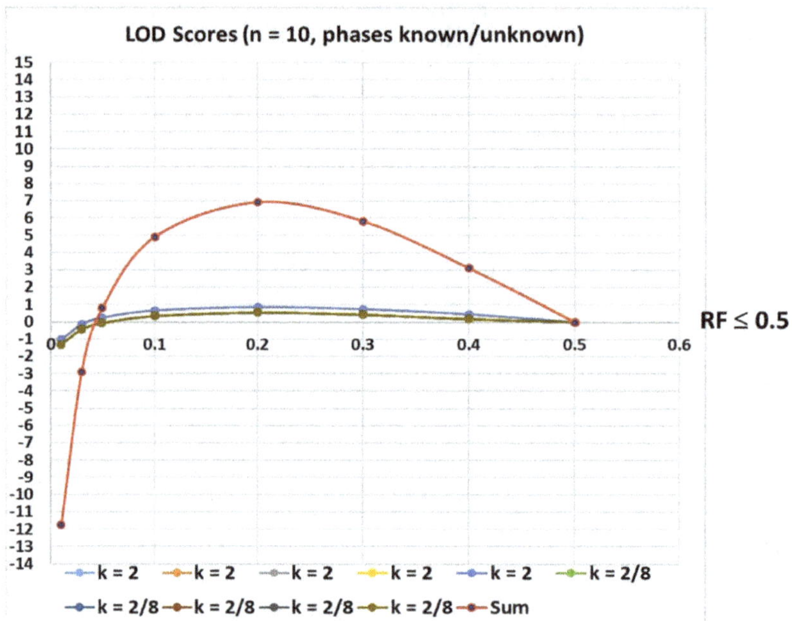

Model n = 10	0.01	0.03	0.05	0.1	0.2	0.3	0.4	0.5
k = 2	-1.02	-0.14	0.23	0.64	0.84	0.73	0.44	0
k = 2	-1.02	-0.14	0.23	0.64	0.84	0.73	0.44	0
k = 2	-1.02	-0.14	0.23	0.64	0.84	0.73	0.44	0
k = 2	-1.02	-0.14	0.23	0.64	0.84	0.73	0.44	0
k = 2	-1.02	-0.14	0.23	0.64	0.84	0.73	0.44	0
k = 2/8	-1.33	-0.44	-0.07	0.34	0.54	0.43	0.18	0
k = 2/8	-1.33	-0.44	-0.07	0.34	0.54	0.43	0.18	0
k = 2/8	-1.33	-0.44	-0.07	0.34	0.54	0.43	0.18	0
k = 2/8	-1.33	-0.44	-0.07	0.34	0.54	0.43	0.18	0
k = 2/8	-1.33	-0.44	-0.07	0.34	0.54	0.43	0.18	0
Sum	-11.75	-2.9	0.8	4.9	6.9	5.8	3.1	0

FIGURE 7-6. The linkage phase has a smaller impact in studies involving multiple families

The curves of LOD scores calculated for linkage phase unknown over the full range of RF (θ) values will appear different from those for linkage phase known. An example of a family of $n = 10$, with either $k = 2$ (blue curve) or $k =$ either 2 or 8 (orange curve) is shown in **Figure 7-5**. Essentially, the phase-unknown curve (orange) is down-shifted over the full range of θ values when compared with the phase-known curve (blue), reducing the power of the linkage analysis in a single family.

It is very interesting to see the results from five families of $k = 2$ (linkage phase known) and another five families of $k = 2$ or 8 (linkage phase unknown) (**Figure 7-6**). The summary LOD scores at RF (θ) = 0.1, 0.2, 0.3, and 0.4 are all above the threshold of 3 for evidence of linkage. In a rather literal interpretation it appears that the LOD score analysis worked the linkage phase out in the unknown situation to dismiss the 8 as the recombinant when it appeared five times out of 10 children. Thus, families with unknown linkage phases are not necessarily automatically uninformative in linkage analysis with LOD scores.

7.3 The Last Place We Looked

The parametric linkage analysis with a resulting LOD score significantly higher than the threshold value of 3 will provide an estimate of the RF (θ) for the allele recombination across the disease D/d and the marker M1/M2 loci. The D/d locus can be on either side of the M1/M1 locus and so additional markers around the first marker on both sides would be employed, one at a time, to refine the location to as narrow a chromosome region as possible. The exercises are repeated with finer marker maps until, ideally, a marker allele presents an association relationship with the disease allele.

The biggest problem at the beginning of the disease gene mapping does not go away with the chromosomal localization of D/d, as we still have no information on what to look for in the DNA sequence in that region. This was especially true in the early examples and additional molecular biology or genetic tricks were required to reveal a "candidate gene" for the disease. But thanks to these tricks and especially the modern variations of them, identifying a candidate gene at a genetic locus has become far less challenging.

We now look at the representative early examples of disease gene mapping by linkage analysis.

7.3.1 HUNTINGTON DISEASE (HD)—THE FIRST AUTOSOMAL DOMINANT DISEASE GENE MAPPED

Mapping of the HD gene was a textbook example of such endeavors. A large five-generation family from the National Research Roster for Huntington's Disease Patients and Families at Indiana University was selected for linkage analysis, along with a larger seven-generation family with interrelated HD allele carriers from a unique community of Lake Maracaibo, Venezuela. An anonymous DNA fragment/probe, that is, with no known information on function, which was reported to detect RFLPs in human populations was used as a genetic marker. An RFLP is a "restriction fragment length polymorphism," resulting from a single-nucleotide change, which either removes or creates a cutting site for a restriction endonuclease. These enzymes are from bacteria and cut DNA at sequence-specific sites. The changes in the sites would present as the DNA fragment length differences and the patterns are used to genotype individuals. Thus, an RFLP essentially detects an SNP. This probe was localized to human chromosome 4 by Southern blot hybridization using human-mouse somatic cell hybrids, carrying a specific single human chromosome in each cell line. To this marker, the HD locus had a combined LOD score of 8.53 at RF (θ) = 0 from the analysis of the two families, that is, odds of >338 million to 1 that the marker locus overlaps with the HD locus.

Follow-up studies narrowed the candidate gene region down to a 500-kbp (kilo base pair) segment between two polymorphic DNA markers D4S180 and D4S182 as the most likely site to contain the genetic defect causing HD in the patients. *Gene bashing* is a term that is no longer used very often today, but quite popular at the time, and describes the final step of identifying the candidate HD gene very well. This can be translated into interrogating a region of genomic DNA asking the question of "does it contain any sequence that is transcribed into an mRNA?". The technique called "exon trapping" (obsolete today) was used to reveal a coding sequence of 9,432 nucleotides, with a predicted protein of 3,144 amino acid residues and a predicted molecular weight of 348 kDa (kilodalton). This initially cloned allele contained a (CAG)n trinucleotide repeat sequence with 21 units, coding for 21 glutamine (Q) residues. In 173 chromosomes tested from normal subjects, there were 17 alleles of varying lengths in the (CAG)n repeat, from 11 to 34. Most (98%) were in the 11–24 range, two were in the 25–30 range, and two were 33 and 34 each. In the American family studied, a normal subject had 36 units, and two HD subjects had 44 and 46 units, respectively. In the Venezuelan family, the normal allele is 33 units and the HD allele is 49 units. This gene is now known as the *HTT* gene, coding for the huntingtin protein, and was "the last place we looked," in terms of mapping the HD disease gene.

7.3.2 CYSTIC FIBROSIS (CF)—THE FIRST AUTOSOMAL RECESSIVE DISEASE GENE MAPPED

Mapping the CF gene also relied on RFLP markers across the human genome. With a large number of such markers studied but no discovery of linkage to CF, the initial efforts at least excluded 40% of the human genome as potential sites of the CF gene. If we use a keychain analogy, on a chain of 100 keys (markers), 40 have been shown to be unable to open the lock (disease gene). In the continuing efforts involving 200 additional RFLP markers, a marker on human chromosome 7 (D0CRI-917) was shown to be linked to the CF locus with a peak LOD score of 3.96 at RF (θ) = 0.14. This genetic distance roughly translates to 15 million base pairs (Mbp) of genomic DNA, either way of the marker so the CF gene is located within a 30-Mbp region, now at about 1% of the human genome. Mapping with additional markers narrowed the region down to between two closely linked markers that flank the CF gene, MET and D7S8 on the long arm of chromosome 7. All the stops were pulled out in the arsenal of tools available for genomic DNA analysis and "gene bashing" to select for the best DNA probes for use to screen cDNA libraries of epithelial cells, the cell type relevant to the disease phenotype.

From the cDNA clones with overlapping sequences a full-length cDNA was assembled to encode a predicted polypeptide of 1,480 amino acids with a molecular weight of 168 kDa. The predicted protein contains two repeated motifs, spanning the membrane several times and resembling the consensus nucleotide (ATP)-binding folds (NBFs), suggesting a function of ion (chloride) transport, consistent with the pathology of the disease. The new putative protein was named cystic fibrosis transmembrane conductance regulator (CFTR). Approximately 70% of CF patients have a specific deletion of three base pairs, the codon of a phenylalanine (Phe/F) residue at position 508 of the putative CFTR protein, ΔF508.

7.3.3 ALKAPTONURIA (AKU)—MAPPING THE GENE FOR THE FIRST HUMAN GENETIC DISORDER TO BE DESCRIBED

We have previously described an outline of the landmark events leading up to the localization of the alkaptonuria disease locus and the identification of the gene, homogentisic acid oxidase (HGD). The breakthrough came about in a fashion not dissimilar to the fruit fly genetic experiments conducted in the Morgan laboratory in the early 1900s, involving all phenotypic loci by a method called "homozygosity mapping." Essentially, this method relies on the appearance of alleles of closely linked loci on the same physical chromosome. Such close genetic distances do not allow the alleles to separate via recombination over relatively short periods of time, in terms of human history. They are said to be in linkage disequilibrium (LD), which is defined as alleles across the loci presenting biased instead of the random combinations. If the closely linked loci are two recessive disease loci, the coinherited disease alleles tend to have a greater chance of ending in a homozygote in consanguineous families, manifesting both recessive diseases in one individual more frequently.

In two consanguineous families presenting a pattern of coinheritance with either neonatal severe hyperparathyroidism (NSHPT) or sucrase-isomaltase deficiency, a chromosome 3 locus potentially harboring the alkaptonuria candidate gene was mapped. Further analysis with polymorphic DNA markers narrowed the candidate alkaptonuria gene to within a 16-cM region at 3q2. A centimorgan (cM) is a genetic map unit of 1% recombinants, equivalent to roughly 1 million base pairs of DNA on average. The combined LOD scores from two consanguineous and two nonconsanguineous families was 4.3 (odds of 19,000 to 1), at RF (θ) = 0. Neonatal severe hyperparathyroidism (NSHPT) is the recessive form of familial hypocalciuric hypercalcemia (FHH), an autosomal dominant condition caused by a gene defect on chromosome 3q21-24. The *FHH* gene was mapped to the 3q2 region with a combined LOD score of 20.67 to the rhodopsin (RHO) gene locus, which was mapped to 3q21-24 first among the three genes mentioned here, but critically by "forward genetics."

Unlike in the cases of the HD and the CF genes, for which the putative coding sequences and predicted protein sequences were deduced from the captured transcribed mRNA sequences, the identification of the human *HGD* gene required a detour by way of the identification of a fungal gene and comparison of its sequence to randomly cloned expressed human sequences. We have looked at the technical details in those efforts in **Chapter 2**.

7.4 Completing the Circle—Koch's Postulates in the Validation of Gene Mapping Results

Koch's postulates were formulated from his work on *Mycobacterium tuberculosis* and are broadly applicable to molecular genetics. We will make an attempt at translating these and outline what remains to be done to complete the circle of disease gene mapping:

- Koch's: A specific microorganism is always associated with a given disease.
  Molecular genetics: An allele of a locus is always associated with a given genetic disorder.
- Koch's: The microorganism can be isolated from the diseased animal and grown in pure culture in the laboratory.
  Molecular genetics: The disease allele can be isolated and cloned (into a vector).
- Koch's: The cultured microbe will cause disease when transferred to a healthy animal.
  Molecular genetics: The cloned disease allele, in the proper genetic context, that is, heterozygote for dominant or homozygote for recessive traits, will reproduce the genetic disorder when transferred into or created in a (healthy) animal model.
- Koch's: The same type of microorganism can be isolated from the newly infected animal.

Molecular genetics: The disease allele transferred to or a normal allele deleted in the model animal to reproduce the genetic disorder should be the only genetic event. This one is a slight departure from Koch's original postulate pertaining to microorganisms.

Essentially, we need to show that the specific mutation is only associated with the specific disease and will only cause that specific disease in the proper genetic context. For the *HTT* gene, two telling mouse models of the dominant HD have been made by imposing the human disease variant allele on the background of the mouse alleles. The first was the R6/2 mouse harboring a transgene consisting of the promoter and exon 1 of the human *HTT* gene, with 150 (CAG)n units. These mice develop rapid and robust behavioral pathologies of the human disease. Another transgenic mouse model (BACHD) used a full-length mutant human huntingtin (mHTT) with a mixed 97 CAG-CAA repeats and its native promoter and control elements. The BACHD mice presented progressive motor deficits, neuronal synaptic dysfunction, and late-onset selective neuropathology. This model also contained an off-switch, known as the *loxP* sites that flanked the engineered mutant exon 1, which can be removed in a tissue/cell type that expresses the *Cre* recombinase. This design would allow the use of this mouse model for in-depth analysis of HD pathology at the cellular level.

For the recessive disease CF, the mouse alleles in the *Cftr* locus was replaced with a ΔF508 allele using a gene-targeting technique known as "knock-in." In the ΔF508 homozygote mice the epithelial cells lacked the CFTR protein in the apical membrane, which became impermeable to chloride ion, recapitulating the human CF abnormalities.

7.5 Linkage Analysis Today

Gene mapping by linkage analysis was an effective method in the early days of identifying disease genes, achieving monumental successes in the examples of the HTT and the CFTR genes, representing the two modes of Mendelian inheritance in human genetic disorders. Today the main force by which disease genes/alleles are discovered is the topic of the next chapter, genome-wide association studies (GWAS). The GWAS asks no questions on the mode of inheritance and operates on the ultimate level of looking for allele sharing, at least by intent. As we will see in the next chapter, the biggest issue with the results of GWAS is the potentially large number of false positives. At the end of the day, it is still a genetic disorder and linkage analysis can still provide a powerful tool for validation of the GWAS findings.

BIBLIOGRAPHY

1. https://scrapsfromtheloft.com/comedy/jeff-foxworthy-totally-committed-transcript/
2. https://www.britannica.com/biography/Robert-Koch
3. A novel gene containing a trinucleotide repeat that is expanded and unstable on Huntington's disease chromosomes. The Huntington's Disease Collaborative Research Group. *Cell.* 1993 Mar 26;72(6):971–83. doi:10.1016/0092-8674(93)90585-e. PMID: 8458085.
4. Bailey-Wilson, Joan E (June 2018) Parametric and nonparametric linkage analysis. In: eLS. Chichester: John Wiley & Sons, Ltd.
5. Gray M, Shirasaki DI, Cepeda C, André VM, et al. Full-length human mutant huntingtin with a stable polyglutamine repeat can elicit progressive and selective neuropathogenesis in BACHD mice. *Journal of Neuroscience.* 2008 Jun 11;28(24):6182–95.
6. Gusella JF, Wexler NS, Conneally PM, Naylor SL, Anderson MA, Tanzi RE, Watkins PC, Ottina K, Wallace MR, Sakaguchi AY, et al. A polymorphic DNA marker genetically linked to Huntington's disease. *Nature.* 1983 Nov 17–23;306(5940):234–8.
7. Knowlton RG, Cohen-Haguenauer O, Van Cong N, Frézal J, Brown VA, Barker D, Braman JC, Schumm JW, Tsui LC, Buchwald M, et al. A polymorphic DNA marker linked to cystic fibrosis is located on chromosome 7. *Nature.* 1985 Nov 28–Dec 4;318(6044):380–2.
8. Kurbasic A, Hössjer O. On computation of p-values in parametric linkage analysis. *Human Heredity.* 2004;57(4):207–19.
9. Nyholt DR. All LODs are not created equal. *American Journal of Human Genetics.* 2000 Aug;67(2):282–8.
10. Paigen K, Petkov PM. PRDM9 and its role in genetic recombination. *Trends Genetics.* 2018 Apr;34(4):291–300.
11. Riordan JR, Rommens JM, Kerem B, Alon N, Rozmahel R, Grzelczak Z, Zielenski J, Lok S, Plavsic N, Chou JL, et al. Identification of the cystic fibrosis gene: Cloning and characterization of complementary DNA. *Science.* 1989 Sep 8;245(4922):1066–73.
12. Rommens JM, Iannuzzi MC, Kerem B, Drumm ML, Melmer G, Dean M, Rozmahel R, Cole JL, Kennedy D, Hidaka N, et al. Identification of the cystic fibrosis gene: Chromosome walking and jumping. *Science.* 1989 Sep 8;245(4922):1059–65.
13. Sparkes RS, Klisak I, Kaufman D, Mohandas T, Tobin AJ, McGinnis JF. Assignment of the rhodopsin gene to human chromosome three, region 3q21-3q24 by in situ hybridization studies. *Current Eye Research.* 1986 Oct;5(10):797–8.
14. Susgun S, Kasan K, Yucesan E. Gene hunting approaches through the combination of linkage analysis with whole-exome sequencing in Mendelian diseases: From Darwin to the present day. *Public Health Genomics.* 2021;24(5–6):207–217.
15. Tsui LC, Buchwald M, Barker D, Braman JC, Knowlton R, Schumm JW, Eiberg H, Mohr J, Kennedy D, Plavsic N, et al. Cystic fibrosis locus defined by a genetically linked polymorphic DNA marker. *Science.* 1985 Nov 29;230(4729):1054–7.
16. Zeiher BG, Eichwald E, Zabner J, Smith JJ, Puga AP, McCray PB Jr, Capecchi MR, Welsh MJ, Thomas KR. A mouse model for the delta F508 allele of cystic fibrosis. *Journal of Clinical Investigation.* 1995 Oct;96(4):2051–64.

8

A Genetic Association Study Is Disequilibrium Linkage Analysis

It does not detect genetic association

In our highlights of "the last places we looked," anonymous DNA fragments/probes detecting restriction fragment length polymorphisms (RFLPs) (SNPs) in human populations were used as genetic markers for linkage analysis with the unknown Huntington disease (HD) locus. One of 12 such anonymous markers was shown to be linked to the unknown HD locus and mapped to human chromosome 4, thus bringing the HD locus to this specific chromosome. This was quite a theme in the embryonic days of gene mapping, repeated in the efforts of mapping the cystic fibrosis (CF) locus, involving a much larger number of such markers. One of these hundreds of markers, located on human chromosome 7, was shown to be linked to the unknown CF locus, now known to be on chromosome 7. We compared the screening of the hundreds of markers in the early steps of CF gene mapping that eliminated 40% of the human genome as ruling out 40 keys on a 100-key keychain for a lock. In every sense, this is a perfect comparison. In a research paper it takes no more than one simple sentence in one line to state how many markers were tested, that is, keys tried. Behind the scenes in the laboratories, it was tried one key at a time.

There was then a tremendous economic burden in these projects, and the burden of the research material: large amounts of the genomic DNA from each subject enrolled in the studies. Typically, the projects required collaboration among many institutions with a large team of investigators and support staff. The HD gene mapping paper in 1983 had 20 coauthors from five institutions, and the CF gene mapping paper in 1985 had 14 coauthors from five institutions. Behind every factor that enabled the successes in the gene mapping of huntingtin (HTT) and cystic fibrosis transmembrane conductance regulator (CFTR), and many others, is a potential challenge or obstacle that the classical parametric linkage analysis may not be able to overcome. Large multigeneration families may not always be available or the disease is very rare.

8.1 Complex Diseases

HD and CF are single-gene or monogenic diseases in which the cause-and-effect path is rather short and direct. In HD, the expansion of the (CAG)n codon for the amino acid glutamine (Gln/Q) results in an expansion of Gln/Q, known as poly-Q, in the polypeptide chain. Including HD there are nine poly-Q neurodegenerative disorders, eight being autosomal dominant and one X-recessive. The extra-long poly-Q in the mutant protein causes abnormal folding and proteolytic degradation of the huntingtin protein, producing toxic poly-Q peptides, which aggregate. There is a loss of function from the normal huntingtin protein, and a "gain" of (bad) function by the toxic aggregated waste to cause neuronal dysfunction and neuronal death. In CF, the mutated CFTR can no longer regulate chloride ion transport, perturbing the fluid balance across the airway epithelium, causing the respiratory symptoms.

From our review of the principles of genetics, discovered by Mendel and put into meaningful practice of genetic mapping by Morgan, we learned that genetics is practiced by looking at only two loci at a time. The monogenic diseases are one of the two and a marker is the other, each time a parametric linkage analysis is performed. Much more challenging would be the complex diseases which are the results of (many) more than one genetic component and additional environmental components (**Figure 8-1**). The disease manifestation is no longer from a single (monogenic) locus, and our system of linkage analysis is thus broken. At the practical level, the reason is that the "recombinants" called by the D/d designation of the disease relative to the marker alleles are no longer true, as D/d is now multiple instead of just one locus. An example of such a disease is the type II diabetes (T2D). According to the National Diabetes Statistics Report by the Centers for Disease Control and Prevention, 38.4 million people (11.6%) of the U.S. population have diabetes, whereas an additional 97.6 million have prediabetes. Included among these are the individuals affected by type I diabetes (T1D), that is, the monogenic diabetes, at up to 3% of all diabetes diagnosed before age 35, with about 300,000 cases in the United States. Thus, the great majority of diabetes patients are of type II, the complex disease. Worldwide, there are 537 million people with diabetes, more than 90% of whom are T2D.

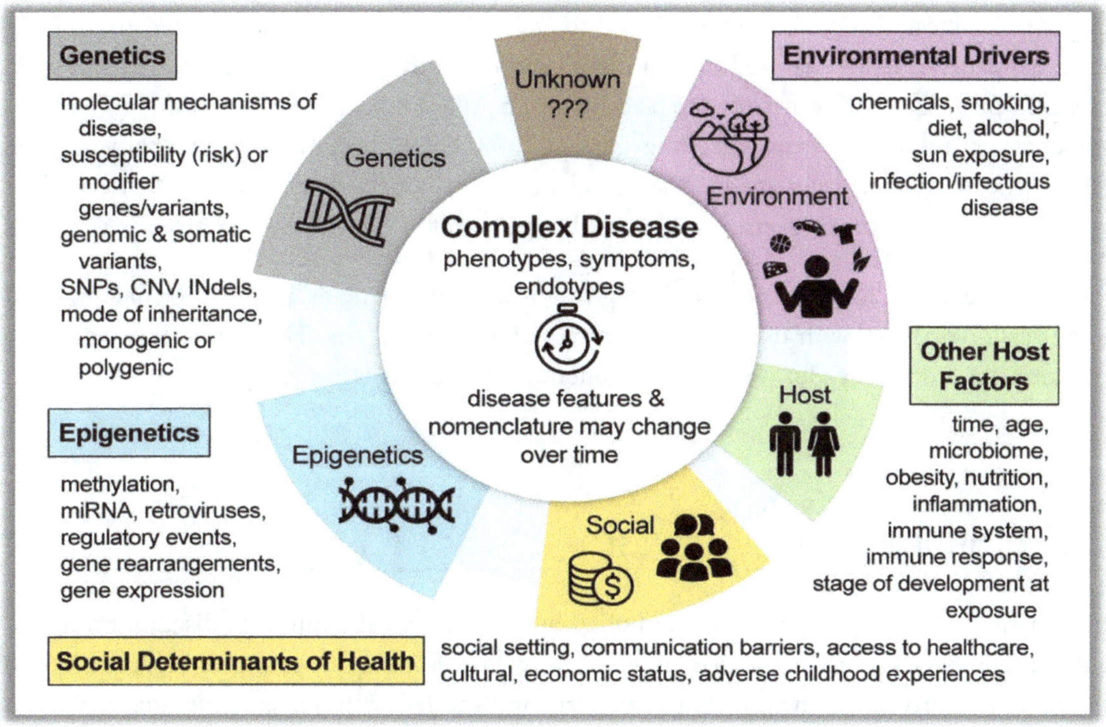

FIGURE 8-1. Complex diseases with genetic and environmental factors

In either case, for monogenic or the complex diabetes, the genetic approach cannot be the same taken for mapping the HD and the CF genes. The "monogenic" designation of T1D can be somewhat misleading. This is referring to the fact that each genetic variant at a single T1D gene locus will directly cause the phenotypic manifestation of diabetes. However, a variant from many T1D loci can produce the same disease manifestation, whereas a true monogenic disease has a one-to-one relationship, such as HD and CF. To date somewhere between 30 and 50 T1D loci have been identified using the "association studies," the topic of this chapter.

So we went for the ultimate directness with something called "association studies." This is similar to the clinical studies of "cases *versus* controls." For example, in order to test if a drug has a beneficial effect on a specific disease, a patient population is recruited and divided into two groups that are as the same as possible by all available criteria, such as age, sex, body mass index, diet, life style, and previous treatment and drug intervention history. The equity of the two groups is to ensure that any potential effect observed is not attributed to any of the criteria considered. One group is given the actual drug, called the treated or treatment group, and the other is given an identical looking but starch pill, called the control or the placebo group. The statistical power required to resolve the differences of "responses" from the two groups guides the determination of how many patients to include in each group. A chi-square test of independence is performed to see if there is a biased distribution between the response levels from the treatment *versus* the control group.

Now, we replace the "high" *versus* "low" drug responses with "diabetes" *versus* "normal" and "treatment" *versus* "placebo" with two allelic genotypes of a marker locus. This is now a genetic (or genomic) association study. But where is the "genetics" in an association study?

8.2 Linkage Disequilibrium

The genetics in an association study is implied by something called linkage disequilibrium (LD). In the discussions of model organism genetics, with Mendel's garden peas and Morgan's fruit flies (at least in the early days), the progenies were kept and studied within a few generations. We would not be able to appreciate the time record marked in the genetic material, the chromosomal DNA, in these short-term experiments. In human genetics, the time history of mutations and mating events in different populations can be recognized and utilized for mapping disease genes.

8.2.1 ALLELES IN A POPULATION

In the subject of "population genetics," founded by the British mathematician G. H. Hardy (1877–1947) with a rather reluctant one-page letter to the editor of *Science* in 1908, the core principle is: in a fairly large evenly sexed population of completely random mating, the allele frequencies of a Mendelian pair of characters within the population established after one generation will not change from and after the second generation. This is known as the Hardy–Weinberg principle, or equilibrium, after Hardy and the German Ob/Gyn doctor Wilhelm Weinberg (1862–1937), who also published his related work independently in 1908. For a two-allele locus A/a, the genotype frequencies at equilibrium are the products of the allele frequencies, p_A and p_a. The genotype frequencies of A/A, A/a, and a/a would be p_A^2, $2p_Ap_a$, and p_a^2, respectively.

Of course the science of population genetics has grown much bigger than the single-page letter from Hardy but much of the science, ironically, is about investigating the factors that would perturb the Hardy–Weinberg equilibrium. For example, human population mating is almost never completely random and probably far from being random when manifested traits are concerned. Population migration can have an impact on the allele frequencies, as well. Further ironically, genotypes that do not bring about manifested traits are mated rather randomly behind the scenes. For example, short tandem repeats (microsatellites) and single-nucleotide variants/polymorphisms (SNVs/SNPs) that are not in the coding or expression regulatory regions of the chromosomal DNA have no way of being selected during mating. To completely kill this joke, even today when whole-genome sequencing (WGS) can report any individual's near total genomic DNA sequence, no one is interested in reading the repetitive DNA or SNP profiles on an online dating site, which would contain more selectable traits or qualities such as height and weight or annual income.

The term equilibrium was used in the Hardy–Weinberg principle to reflect the effects of completely random mating on the population allele frequencies in one locus. Now we add the term linkage, which describes the relationship of two loci closely located on the same chromosome that the alleles across the two loci cannot display a Mendelian recombination rate of 50% (RF = 0.5). That was the core foundation of the linkage analysis applied in disease gene mapping discussed in **Chapter 7**. The genetic linkage, in a model organism, can be detected in one generation of test mating, as we have seen previously. In humans, the LOD score analysis provides us with an answer.

8.2.2 GENOTYPE FREQUENCIES ACROSS TWO LOCI

If we have two loci, regardless of their chromosome relations, that is, whether they are on the same or not and genetically linked or not, with allele frequencies A/a = 0.8/0.2 and B/b = 0.6/0.4. Both loci are at Hardy–Weinberg equilibrium. The expected frequencies of the four different allele combinations across the two loci would be the simple products of the corresponding allele frequencies (**Table 8-1**). For example, in a randomly mating population the chance of getting an "A" is 0.8 and getting a "B" is 0.6. The chance of having the "AB" genotype is $0.8 \times 0.6 = 0.48$.

TABLE 8-1. Expected *versus* observed genotype frequencies across two loci

Genotype	AB	Ab	aB	ab
Expected	0.48	0.32	0.12	0.08
Observed	0.56	0.24	0.04	0.16

Another important condition, not relevant in Mendel's or Morgan's experiments, is the amount of time, or the number of generations required to see a two-locus genotype frequency as the product of the two allele frequencies. This condition is usually met for alleles that had risen a long time ago, but not for relatively recent alleles. For the example given above, the actual two-locus genotype frequency shown in the green-shaded row in **Table 8-1** can be different from the predicted frequency by multiplying the allele frequencies. It would appear that some alleles have a tendency to stay together across the two loci, such as "AB" and "ab," with higher than expected genotype frequencies by a margin of 0.08. The recombinant genotypes fell less than the expected frequencies in a corresponding manner by the same margin of 0.08. This phenomenon is called "linkage disequilibrium," literally the perturbation of the expected genotype frequency by allele frequencies at an equilibrium due to the genetic linkage of the two loci, hence the observed disequilibrium. It is important to note that the somewhat unfortunate definition of this phenomenon as "linkage disequilibrium" says more about our desire for the disequilibrium to be caused by genetic linkage. But the phenomenon is not necessarily caused by genetic linkage.

8.2.3 UNLINKED LOCI REACH "LINKAGE EQUILIBRIUM" FASTER THAN LINKED LOCI

We "go back in history" in a defined human population to the point when a locus A/a was just "A" (technically not a locus without another allele), but the locus B/b already had allele frequencies of 0.6/0.4 at Hardy–Weinberg equilibrium (**Figure 8-2**). The genotype frequency of "AB" would have been 0.6, and 0.4 for "Ab." When a new variant/mutation "a" occurred, "A" became A/a. The new variant "a" could be on the same chromosome with "B" or "b," and let's assign it to "b." At this moment in time, there were only three genotypes across the two loci: "AB," "Ab," and "ab," but no "aB." The genotype frequencies would have been 0.6 (AB), 0.4 (Ab), and nearly negligible (ab), respectively, if the population is large enough. The allele frequency for "a" is 1/(2*N*), with the population size number of *N*. And if $N = 100{,}000$, $p_a = 0.000005 = p_{ab}$.

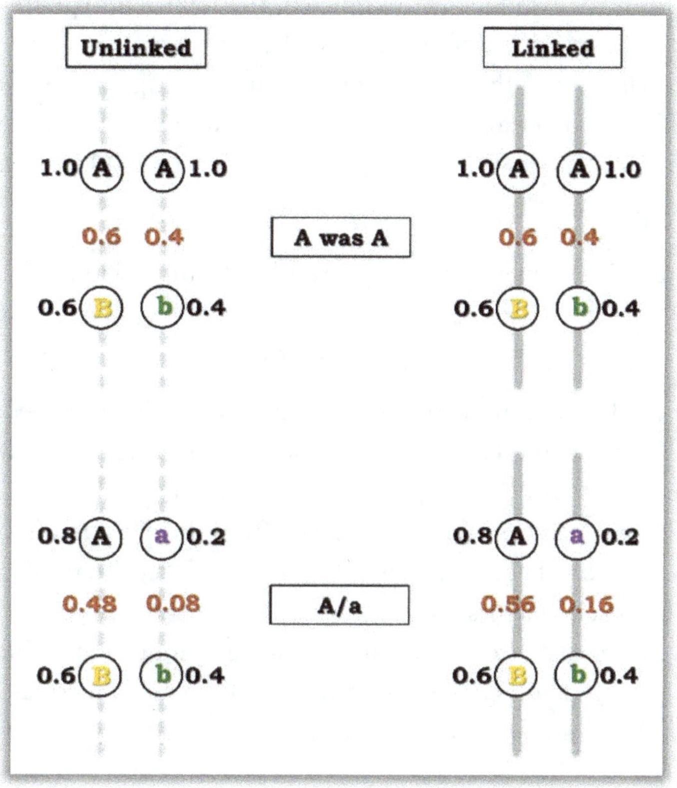

FIGURE 8-2. Unlinked loci reach "linkage equilibrium" faster than linked loci

Most of the time, a new mutation will disappear because of its rarity. But humans are estimated to generate ~70 single-nucleotide mutations *per* generation for a diploid genome, with ~2.2 being deleterious and ~0.35 resulting in a change of amino acid. The majority of the mutations are produced in the male germ cells. Some of these new mutations, however, can be amplified by a selective advantage. The "selective advantage" is not necessarily, and often not, at the organismal level, but at the cellular level in the primordial germ cells (PGCs). The molecular mechanisms for the selective advantage remain mostly unknown but perceived from the facts of allele amplification.

This happens in one person; the amplification of the new minor allele would not have an effect on the population level. To affect the population, the new allele has to be successful in reproduction and continued mating events. Overcoming all obstacles of population genetics, the minor allele frequency can be increased over time to reach a much more significant level. At the time when the minor allele at A/a attained the population frequency of 0.2 as seen in our example, at Hardy–

Weinberg equilibrium, if the two loci were unlinked, the four genotype frequencies would be at their maximum, that is, the products of the individual allele frequencies. The new "aB" genotype, at a frequency $0.2 \times 0.6 = 0.12$, took this amount down from the original "AB" genotype frequency of 0.6, making it 0.48. If the two loci are linked, the alleles cannot recombine at the maximum rate of 0.5 *per* generation, and the takedown of "AB" frequency is less than 0.12 across the same amount of time, at 0.04 in our example.

Linkage disequilibrium (LD) is mathematically formulated simply with:

$$D_{AB} = p_{AB} - p_A p_B$$

D_{AB} is the coefficient of LD between the two loci A/a and B/b, defined as the difference between the frequency of the "AB" genotype and the product of the "A" and "B" allele frequencies. For our example with the two bi-allelic loci, $D_{AB} = -D_{Ab} = -D_{aB} = D_{ab}$. Mathematically, the "AB" genotype under LD is often expressed as a "gain" of frequency by D_{AB} from the expected frequency of $p_A p_B$. Quite literally, the expected is 0.48 but the observed is 0.56, a gain of 0.08. To think of this as a gain would imply that there are forces and factors that preferentially associated the two alleles together. In fact, it is not a gain but a longer distance to the point of linkage equilibrium, which requires and is the measurement of the passage of time. This inverted view of the meaning of D_{AB} is more accurate for the understanding of the conditions that create LD.

8.2.4 "LINKAGE DISEQUILIBRIUM" DOES NOT REQUIRE LINKAGE

Another mathematical aspect of LD is reminiscent of the one-page Hardy letter that defined the framework of population genetics. These concepts measure the differences in numbers without having to have some meaningful examples in biology. From this perspective, a value of D_{AB} is simply an observed difference. We call it the detection of LD but this does not require linkage. The one generation of random mating to fix the population allele frequencies in the Hardy–Weinberg equilibrium is for the alleles of one locus. Even for two unlinked loci, including two on different chromosomes, the maximum recombination rate is only 0.5 in one generation, a major compromise to the complete randomness required for reaching equilibrium of allele combinations across the two loci. For two linked loci, the recombination rate is less than 0.5, the closer the linkage the lesser the rate. Thus two unlinked loci reach the point of "linkage" equilibrium faster than if the two loci were linked. Over time, the mathematical distances covered toward the eventual equilibrium of the four genotype frequencies will show a difference for us to detect. In essence, a genetic association study is in principle still linkage analysis, but we can call it "disequilibrium linkage analysis." It is not performed using families but involving unrelated individuals to determine the degree of "disease allele sharing" among them *versus* among the normal control subjects who are also unrelated. Later we will see that genome-wide association studies (GWAS) tend to produce a large number of false positive findings, which are very much the detected "linkage disequilibrium" between two loci without an actual genetic linkage. Many factors other than purely genetic linkage can produce the biased allele combinations/distributions across the two loci. We are testing for LD at the present time point without the benefit of knowing when "A" became A/a.

8.3 Genetic (Genomic) Association Studies

8.3.1 THE ABSOLUTE ASSOCIATION BETWEEN A VARIANT ALLELE AND A PHENOTYPE

The luckiest keychain to lock trial would be for the first key to be tried to open the lock. We will use an ideal example, such as the A to T mutation that created the sickle cell disease (SCD). Let's pretend that the genetic information did not come from "forward genetics" via the analysis of the mutant β-globin protein and the ridiculously abundant β-globin mRNA in the rabbit reticulocytes. Instead, we "somehow" happened upon this pair of SNPs located on human chromosome 11 at 11p15.4, with allele frequencies A/T = 0.9/0.1 in a population in which SCD is prevalent. Still "somehow" we decided to see if there is a special relationship between the A/T SNPs and the normal/SCD phenotypes of unknown genetic basis. In an idealistic model genetic association study involving 40 individuals in the SCD group (cases) and another 40 in the Normal group (controls), we performed genotyping and determined the numbers in the cases and the controls for each of the three possible genotypes, T/T, T/A, and A/A (**Figure 8-3**). Individuals with the T/A and A/A genotypes were clinically normal, and individuals with the T/T genotype were diagnosed with SCD. A chi-square analysis validated the association, with a chi-square of 74.18, relative risk (RR) of infinity, 95% CI (confidence interval): 2.04–100, $p < 0.05$. We would have a high degree of confidence to claim that the T/T genotype is the genetic determinant of the recessive SCD. The 100% T/T to SCD relationship is the strictly defined "genetic association." We know for a fact that this is true, and this is actually the basis of the very first molecular diagnostic test for a genetic disorder. Such a test would require the 100% association.

Genotypes/ Phenotypes	SCD	Normal	Total
T/T	40	0	40
T/A	0	8	8
A/A	0	32	32
Total	40	40	80

Chi-square = 74.18, *d.f.* = 2, *p* < 0.05 (= 7.79e-17)
Relative risk: ∞ (comparing T/T to A/A)
95% CI: 2.04 – 100

FIGURE 8-3. Chi-square analysis on an absolute genetic association with the phenotype

8.3.2 THE DEGREES OF ASSOCIATION BETWEEN A VARIANT ALLELE AND A PHENOTYPE

Now we look at another randomly located SNP site with the alleles A/T at frequencies 0.6/0.4 and ask the question if there is an associative relationship with another recessive "Disease." This time, we involved 90 individuals that gave us the genotype and case-*versus*-control distributions as shown in **Figure 8-4**. In this example, a chi-square analysis still indicated an association, with a chi-square of 10.20, relative risk (RR) of 7.14, 95% CI: 0.32–1.05, $p < 0.05$. We can reject the null hypothesis that the Disease/Normal (D/N) phenotypes and the T/T, T/A, and A/A genotypes are independent. How useful is the result of this test? We can say that the T/T genotype must not be causal to the disease because there were individuals with the A/A genotype in the Disease group. But is this A/T SNP locus linked to the D/N locus?

Genotypes/ Phenotypes	Disease	Normal	Total
T/T	22	7	29
T/A	16	22	38
A/A	7	16	23
Total	45	45	90

Chi-square = 10.20, *d.f.* = 2, *p* < 0.05 (= 0.006)
Relative risk: 7.14 (comparing T/T to A/A)
95% CI: 0.32 – 1.05

FIGURE 8-4. Chi-square analysis on genetic association with the phenotype

8.4 Whose Favorite (Candidate) Gene?

In the second example above, we emphasized on the SNP locus being "randomly located" in the human genome. An interpretive approach to genetic association studies is called the "candidate gene association studies" (CGAS). In many cases, the "candidate" gene becomes the researcher's "favorite gene." By prior knowledge of the gene and the gene's protein product in terms of its biochemistry and the interactive partners in a biological system or pathway, an "educated" guess would be made toward a potential mechanistic involvement in a disease of yet unknown genetic basis. This can be a single-gene disorder, or a complex multigenic one, for which the "candidate" contributes to a portion of the mechanism or a subset of the disease population. Most of the time, by "prior knowledge" we refer to genes that were cloned and sequenced by "forward genetics."

8.4.1 CANDIDATE GENES FOR A COMPLEX DISEASE—ESSENTIAL HYPERTENSION

The kallikrein-kinin system (KKS) was discovered in the early 1920s, starting with the injection of human urine into dogs by the German surgeon Emil-Karl Frey (1888–1977), who observed a marked reduction in the arterial blood pressure as a result. The blood pressure reduction was caused by an "F-substance," speculated by Frey as a peptide hormone produced in the pancreas and excreted in the kidney. The F-substance was named as "kallikrein," from the Greek word for pancreas, kallikreas, and identified a decade later as a proteolytic enzyme. Kallikrein releases the vasoactive peptide "DK"/kallidin (lys-bradykinin) from a plasma protein kininogen, produced in the liver. Lys-bradykinin and its proteolytic derivatives engage bradykinin receptors (BDKRB1 and BDKRB2) on the blood vessel endothelial cells to lower the blood pressure. The KKS counteracts against the renin-angiotensin system (RAS). Renin is an aspartic protease produced in and released from the kidney into the circulation to convert angiotensinogen to angiotensin I, which is further converted to angiotensin II by the metalloprotease angiotensin-converting enzyme (ACE). Angiotensin II increases the blood pressure after engaging the G protein-coupled angiotensin II receptors (ATR1 and ATR2).

The "human urinary kallikrein" (HUK) in Frey's dog experiments is also referred to as the tissue kallikrein, in contrast to the plasma kallikrein, which is produced in the liver and secreted into the blood. The kallikreins are serine proteases that use the hydroxyl group in the active-site serine residue's side chain to catalyze the proteolysis, with varying degrees of substrate specificity. There are three members in the human tissue kallikrein family, the HUK, now KLK1, KLK2, and KLK3, *aka*, the prostate-specific antigen (PSA). In the early 1990s, an HUK-specific serine protease inhibitor (serpin) was identified and its gene was cloned and sequenced. This serpin-class inhibitor of HUK is named kallistatin (SERPINA4).

The human genes of the KKS and RAS were cloned by means of classical forward genetics, or variations such as expression cloning, or homology cloning. In other words, none was by genetic means, such as linkage analysis. The biochemistry, endocrinology, physiology, and pathophysiology of the components in these systems were well studied. Meanwhile, heart disease is the leading cause of death worldwide with hypertension being the main risk factor, affecting ~1.4 billion adults. For up to 90% of the hypertensive patients there is no identifiable specific cause, and their condition is referred to as "essential" or "primary" hypertension, which is now considered a syndrome of "secondary" subsets of endocrine malfunctions. It is estimated that 25%–60% of the general population have a heritable form of the condition, with significant interplays between genetics and the environment.

Two things stand out from the very brief introduction of the KKS and the RAS, which regulate the blood pressure in opposing ways. First, in such complex and intricate multiplayer systems that also "crosstalk," a perturbation at any point would have an expected impact and outcome at the phenotypic level of blood pressure change. Thus, hypertension is truly a perfect example of a complex (hereditary) disease involving potentially many genes and their protein products, and many environmental factors that can act at the gene, as well as the protein level. Second, the overall manifested phenotype of high blood pressure can be potentially deconvoluted by the specific pathways in which the components malfunctioned to produce the phenotype. For example, about half of the disease burden in hypertension is induced by high dietary salt intake (**Figure 8-5**). Thus, the mean arterial pressure (MAP) change (ΔBP) in response to dietary salt can be used to stratify the subjects, with 3–5 mmHg as "normotensive salt-sensitive," 8–10 mmHg as "hypertensive salt-sensitive," and below 3–5 mmHg as "salt-resistant." Salt deprivation can cause a larger drop of MAP in the salt-sensitive subjects than in the salt-resistant or the normotensive salt-sensitive.

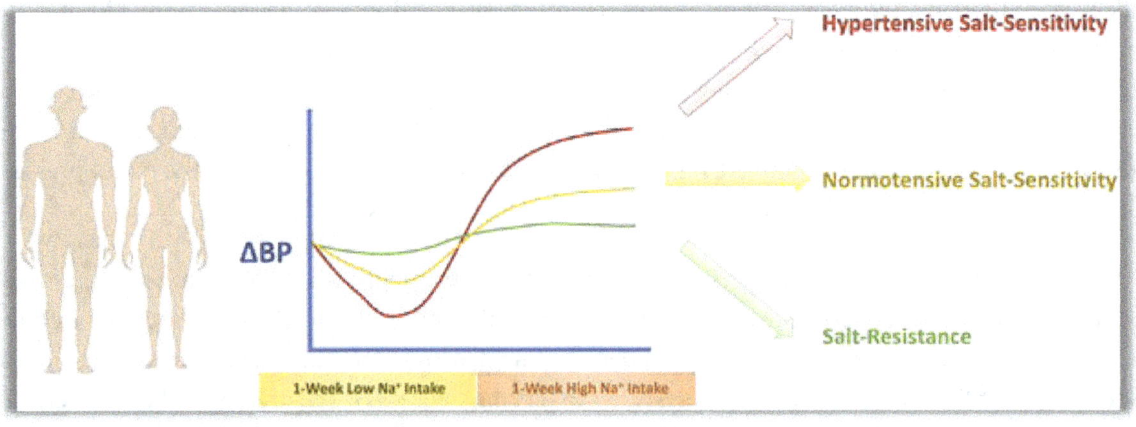

FIGURE 8-5. Salt-induced hypertension

8.4.2 ASSOCIATION STUDIES WITH CANDIDATE GENES IN ESSENTIAL HYPERTENSION

The word candidate is now perfectly suited for the genes in the KKS and the RAS to be considered as the potential genetic determinants of essential hypertension. A study was conducted by The GenSalt Collaborative Research Group in the Han Chinese population involving 1,906 individuals to assess association of salt-induced hypertension with the common genetic variants in 11 candidate genes from the KKS: bradykinin receptor B1 (BDKRB1), bradykinin receptor B2 gene (BDKRB2), carboxypeptidase M (CPM), carboxypeptidase N, polypeptide 1 (CPN1), carboxypeptidase N, polypeptide 2 (CPN2), endothelin converting enzyme 1 (ECE1), kallikrein 1 (KLK1); plasma kallikrein (KLKB1), kininogen 1 (KNG1), the membrane metalloendopeptidase (MME), and kallistatin (SERPINA4).

The SNP rs11847625 in BDKRB2 is bi-allelic (G/C) with the minor allele "C" at a frequency of 0.2. The minor allele "C" was reported by this study to be associated with the MAP increase in response to a seven-day high-sodium diet (18 g of sodium chloride *per* day) administered after a seven-day low-sodium diet (3 g of sodium chloride *per* day). The p value of the association was 1.4×10^{-4}, far less than the 0.05 for falsely rejecting the null hypothesis of no association. What does this finding mean in terms of understanding the genetic basis of a specific subset of hypertension involving one specific gene in a major blood pressure regulation system? The summary statement in the original 2012 report reads like this: "we identified multiple genetic variants of the KKS that contributed significantly to the salt sensitivity of BP" (including seven SNPs in the *ECE1* gene). But, do they?

8.4.3 BONFERRONI—MULTIPLE TESTING CORRECTIONS

What if rs11847625 was tested against the salt-induced hypertension without the knowledge of it being located in the *BDKRB2* gene? In other words, we take away the candidate in the candidate gene. Or, most appropriately, we take away any connection (avoiding the use of either linkage or association here) with the knowledge of a functional gene or protein that has a confirmed biochemical or physiological role in blood pressure regulation. Then it would become a totally random SNP. How good do we or should we feel about the finding, now?

The 2012 report did include another piece of technical information, a very important one, the Q value of the test for "multiple testing corrections." In plain language, if the entire human genome contains only one SNP, and it happens to be rs11847625 and the result would have been conclusive with the reporting of the impressive p value at 1.4×10^{-4}. But there are now about 1 billion reference SNPs in the dbSNP database at the National Center for Biotechnology Information (NCBI),

National Library of Medicine (NLM). This database collects human genomic variations including single nucleotide variations, microsatellites, and small-scale insertions and deletions. Greater than 96% of the human genomic variations are SNPs for which the minor allele frequency is greater than 1%, and most (~99.68%) are bi-allelic.

Now we have a multiple testing situation that any of the SNPs scattered around the human genome could be associated with the phenotype of salt-induced hypertension. For each individual test, a *p* value of 0.05 gives the risk of 5% at falsely rejecting the null hypothesis of no association. For multiple independent tests, this risk will increase significantly. The probability of calling at least one false positive by chance in N number of multiple testing is calculated by $1-0.95^N$. With 20 independent tests, the chance is as high as 64%. Multiple testing correction is performed by setting the *p* value of 0.05 for all the tests performed. Each test now has a *p* value of $0.05/N$. This is called a Bonferroni correction, after the Italian mathematician Carlo Emilio Bonferroni (1892–1960), and is the most stringent at reducing the risk of false positive findings in multiple testing. With the Bonferroni correction, the probability of calling at least one false positive in 20 independent tests is now $1-(1-0.05/20)^{20} = 1-0.9975^{20} = 0.0488$, less than 5%.

8.4.4 FALSE DISCOVERY RATE (FDR)—A STEP BACK FROM BONFERRONI

Testing only rs11847625 against the salt-induced hypertension phenotype does not relieve the burden of multiple testing. We cannot pretend that all the other SNPs do not exist. But if we use the total number of SNPs for Bonferroni correction, the *p* value would be extremely small and the sample size would need to increase to levels impossible for any genetic association study. Not all SNPs would be independent tests toward a disease phenotype, but we have no way of estimating exactly how many, except for those in confirmed complete linkage disequilibrium. For GWAS, it is often corrected by 1×10^6. For candidate gene association studies, a compromise known as the False Discovery Rate (FDR) is commonly used, with the reporting of a Q value: the expected proportion of significant tests resulting in false positives. The Q value for the KKS candidate gene association with the salt-induced hypertension phenotype was reported as 0.03 in the 2012 paper.

8.4.5 CANDIDATES STILL UNCONFIRMED

In the literature, even today, there are volumes of published reports on the association of a variant in "someone's favorite gene," usually the minor allele of an SNP to a phenotype in which the involvement of that gene's protein product "makes the perfect biological sense." The rs11847625 SNP in the *BDKRB2* gene is such an example. Bradykinin is a vasodilator and reduces the blood pressure via engaging its receptors, one of which is BDKRB2. A variant form of the BDKRB2 protein could affect the actions of bradykinin, and thus have a phenotypic output in blood pressure regulation. Alternatively, the associated SNP is not located in the coding region of a protein's gene but in the intronic or regulatory regions and potentially affects the transcription of the gene. The rs11847625 SNP is indeed located in an intronic region, making a functional interpretation more challenging. The authors of the 2012 paper employed the National Institute of Environmental Health Sciences SNP Function Prediction Tool to show that "rs11847625 is located at a transcription factor binding site of BDKRB2, suggesting that this variant could affect the level, location, or timing of gene expression." The authors did concede that "experimental studies will be needed to confirm this inferred functional mechanism."

It has been more than a decade since the association of rs11847625 with salt-induced hypertension was discovered by the candidate gene association study in 2012. There has not been any reported experimental study testing this inferred functional mechanism. This discovery is still actively cited in the recent literature, but the strength of evidence is rated "fair" in a down-escalating scale of "very strong, strong, moderate, and fair."

8.5 The Caveats in CGAS

The first thing we need to understand about the result of an association study on a candidate gene is that we are not that lucky to happen upon the right key for the lock in a 100-key keychain. The following factors (not inclusive) must be considered before the data can be properly interpreted.

8.5.1 ALLELE FREQUENCY AND EFFECT SIZE

For a monogenic disorder the disease (minor) allele contributes the total effect. There are several hundred β-globin gene variants with some causing life-threatening illness such as the SCD. The most well-known variant is an SNP rs334, at chr11:5227002 (GRCh38.p14), the coding sequence change is "A" to "T" (minor allele), changing a Glu to a Val in the sixth position (E6V) of the mature β-globin protein. Globally with a sample size of 6,404 surveyed by the 1000Genomes_30x project, the minor allele frequency is 0.0287, about 3%. It is much more frequent in Africa, at 10% (sample size 1,786), but at zero in Europe (sample size 1,266), South Asia (sample size 1,202), and East Asia (sample size 1,170). The effect of the minor allele can be considered all or none when present or not. In the heterozygote condition, it is called the sickle cell trait (SCT), essentially haploinsufficiency, or conditional haploinsufficiency under stress.

For a multigenic/complex disease, each susceptibility gene is not expected to have such a high level of effect because that would simply qualify that gene's variants for a monogenic disease. More common variants produce low levels of effect, whereas low-frequency variants produce the intermediate to modest levels of effect (**Figure 8-6**).

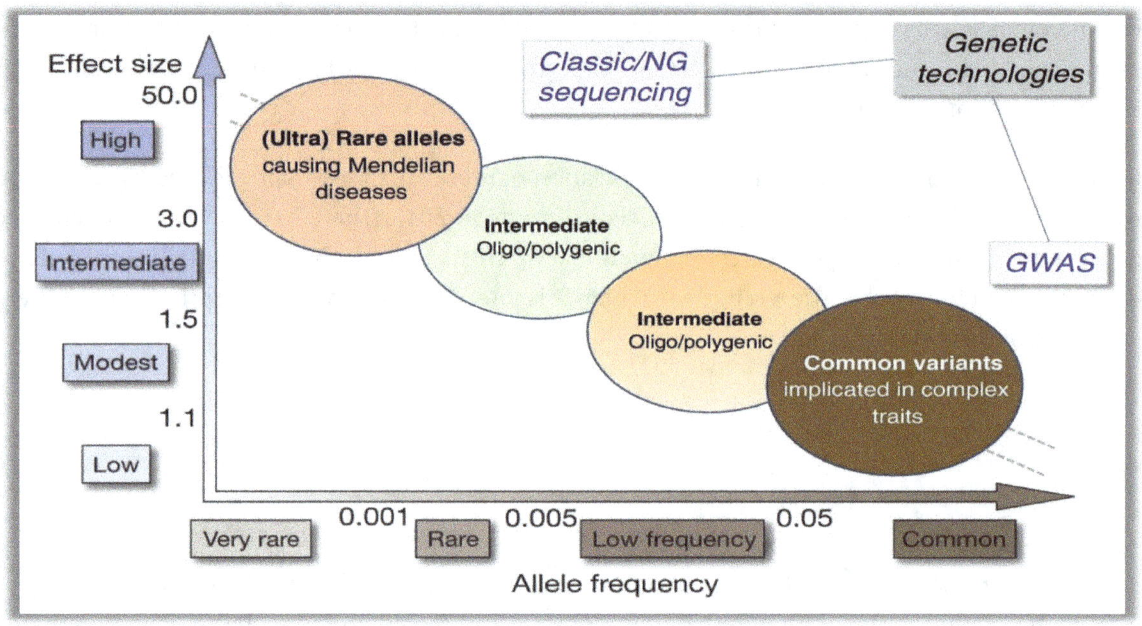

FIGURE 8-6. Allele frequency and effect size in complex diseases

8.5.2 CGAS DOES NOT PROVE ASSOCIATION

Despite the name CGAS, the result is not the strict genetic association, as in the case of A→T at rs334 with E6V to Normal/SCD. That association is 100%, "A" with E-Normal, and "T" with V-SCD, and there is no "A" with V-SCD or "T" with E-Normal in the homozygotes.

The BDKRB2 SNP rs11847625 is located at chr14:96219959 (GRCh38.p14), with an updated minor "C" allele frequency of 24.63% in the global population and a nearly identical frequency of 24.62% in East Asians from the 1000Genomes_30x project. The supplemental data from the 2012 paper are no longer accessible online but surely all four relationships existed for the homozygotes (**Table 8-2**).

TABLE 8-2. Nonabsolute "association" of marker alleles and phenotypes

Allele/Phenotype	Salt-Induced HBP	No Phenotype
G/G	Yes, lesser proportions	Yes, more proportions
C/C	Yes, more proportions	Yes, lesser proportions

There is raw data from another study conducted in 2012 on the association of a ±9 bp (base pairs) polymorphism (dimorphism) in Exon 1 of the *BDKRB2* gene with essential hypertension. The data were described to support association of the +9/−9 genotype with high disease risk against the pooled frequencies of homozygotes (odds ratio (OR) = 1.63, 95% CI = 1.12–2.38, $p = 0.02$), whereas a protective effect was observed for the −9/−9 genotype against disease (OR = 0.59, 95% CI = 0.41–0.86, $p = 0.01$). The ±9 bp polymorphism was hypothesized to have an effect on the rate of gene transcription, as a potential functional mechanism in the disease phenotype. It is important to note that for all three genotypes, −9/−9, +9/−9, and +9/+9, there were both patients and controls, as shown in **Table 8-3**. The "association" is not 100%, and realistically, no mechanistic explanation should be attempted on the effect of the genotype "associated" to the disease phenotype.

TABLE 8-3. Nonabsolute "association" of marker alleles and phenotypes (from Bhupatiraju et al., 2012)

	Patients		Controls	
Genotype	Number	Percentage	Number	Percentage
−9/−9	87	46.1	129	54.7
+9/−9	101	48.3	86	36.4
+9/+9	21	10.0	21	8.9

8.5.3 WE MUST NOT FORGET THE GENETICS IN CGAS

Now we have seen the thought process and actual examples of CGAS for a complex disease in essential hypertension or its subsets. We emphasize that for almost all variants used in such a study the intent and study design were never to test the hypothesis that the minor allele is a mechanistic cause of the disease. The terms "candidate gene" and "association" can be psychologically misleading but we should not let them be scientifically misleading. A common misconception of such association findings is that there is a mechanistic relationship between the tested variants and the disease. It is worth repeating that between the classical linkage analysis and the association analysis/study, the principle did not change: we still rely on linkage and the detection of linkage to map the disease gene. What changed is how the game is played. Instead of relying on large family pedigrees to test the linkage directly, we rely on the population and the passage of time to tell us how linkage prevented the random combination of alleles across the unknown disease locus and a known marker locus.

There is an interesting twist in the published works in genetic association studies and we can use it to drive home the idea of association studies not testing for association. The word "disease" itself should have the implied meaning of being in the minority, as it makes no sense for more than half of the population to walk around with a "disease" whereas the minority is the normal. Intuitively then, the minor allele at the beginning of an association study is often thought to be the risk allele, an appropriate term without a direct mechanistic insinuation. And it is true that most of the time, the minor allele turns out to be the risk allele for the disease. But some minor alleles were found to be associated with a protective effect on the disease model being tested. How do we reconcile this? First, now we should be fully convinced that a genetic association study does not test for association but tests for linkage by disequilibrium. Once that is clear, the explanation is simple and easy. In **Figure 8-2**, we designate A/a as the disease locus and "a" the causative allele, so "A" is Normal and "a" Disease. The B/b locus with "b" being the minor allele is the marker locus in the association study. The linkage disequilibrium gives us an associative relationship between "A" and "B" and also "a" and "b," with "b" the minor allele appearing as the risk allele. But when we made the "a" allele it was randomly assigned to the "b" side. There is no reason that "a" could not have happened on the "B" side, the major allele of the marker locus. An association study in this case would have identified the minor allele "b" of the marker locus, in association with "A," as a protective, instead of a risk, allele.

8.6 Genome-Wide Association Studies (GWAS)

A GWAS is in essence not at all different from the CGAS that we had extensively discussed above. Everything we had learned from CGAS is applicable to GWAS, which is simply to actually perform the large numbers of multiple testing all at the same time, thus, genome-wide. In other words, we stopped pretending that outside of the loci within our favorite genes, all other loci do not exist.

8.6.1 THE FIRST GWAS

The first GWAS in the world was a study to look for genetic variants underlying myocardial infarction, by a collaborative team in Japan, published in November, 2002, in Nature Genetics. This was a perfect transition from the CGAS to a GWAS: genotyping 94 individuals of Japanese descent over 92,788 "gene-based" SNPs. The allele frequencies were then compared with those from 658 individuals from the general population in Japan. A data cleanup was performed, typically involving removal of genotyping errors, deviations from the Hardy–Weinberg equilibrium, and rare minor alleles. The results from 65,671 SNPs (~71%) were retained. A p value of 0.01 was set for association and 99% of the results were excluded. The exclusion was validated using a larger replication panel of 1,133 myocardial infarction subjects and two control groups, 1,006 + 872 (age-matched). An SNP rs909253 (252A→G) in intron 1 of the lymphotoxin-α (*LTA*) gene on chromosome 6p21 showed association with myocardial infarction in the first step, in a recessive model (GG vs. AA + AG) with $\chi^2 = 9.4$, $p = 0.0022$. The replication panel boosted the association to $\chi^2 = 18.0$, $p = 0.000022$, odds ratio = 1.69, 95% confidence interval = 1.32 – 2.15 (against control group 1 of 1,006); and $\chi^2 = 17.4$, $p = 0.000018$, odds ratio = 1.75, 95% confidence interval = 1.35–2.26 (against control group 2 of 872).

This LTA SNP has the allele frequencies of A/G = 0.62/0.34 from the genotyping data of the two control groups. We can view this as the B/b locus in **Figure 8-2**, the genetic marker for the association study. The result at this point can be interpreted as: a myocardial infarction susceptibility gene, A/a, is located in genetic linkage to the LTA SNP rs909253. The "G" allele frequency in the myocardial infarction group is 0.43, presenting clear linkage disequilibrium. The next logical step would be to make an attempt at mapping the A/a locus to a more precise location. In linkage analysis, an initial hit with a marker of greater than the threshold LOD score (3) for an RF value is followed up with more markers near and around the first hit marker. In one direction, a new marker is expected to have a greater than the threshold LOD score for a smaller RF, bringing the mapping closer to the disease locus.

In association studies, the principle of operation is the same. Advances in molecular biotechnology had given us the tools to determine the genotypes along an individual chromosome, called a haplotype, literally meaning the linear genotypes on one half side of the diploid genome in a specific chromosomal region. Haplotypes with linkage disequilibrium will appear as a "fixed block." By a technique called linkage-disequilibrium mapping the unknown myocardial infarction susceptibility gene/locus, A/a was localized to between the genes *P5-1* (MHC class I region ORF) and *AIF1* (allograft inflammatory factor 1).

In an independent study published in 2006, the LTA SNP rs909253 was tested in Japanese and Korean populations for associations with myocardial infarction but the findings were negative. However, the recessive homozygote G/G genotype was significantly associated with the severity of the disease for both test populations. In a meta-analysis of 15 studies with a total of 22,549 myocardial infarction subjects and 16,105 healthy controls, the LTA SNP rs909253 was found to confer a significantly increased susceptibility to myocardial infarction only in Asians.

8.6.2 THE FIRST TRUE GWAS

The first true GWAS in the world was a 2005 landmark paper by a collaborative team of scientists from multiple institutions in the United States, looking for associations between SNPs and age-related macular degeneration (AMD). Some recent review articles counting the number of years past in the history of GWAS regarded this paper as the first GWAS, but the authors of this paper graciously acknowledged the 2002 paper by Ozaki *et al.* as the first GWAS, referred to as "whole-genome association study" at the time.

In the 2005 AMD GWAS paper, 96 cases and 50 controls were carefully selected, including the consideration of potential ethnic bias of the allele frequencies. A true Bonferroni correction was performed using the total number of SNPs for the multiple testing correction of each SNP's association statistics. A total of 116,204 SNPs were genotyped, with 105,980 passing the quality control criteria, including a greater than 85% "call rate" for each SNP in the study subjects and the removal of SNPs with missing alleles or missing homozygotes, or deviating from the Hardy–Weinberg equilibrium. A total of 103,611 SNPs on the autosomes were analyzed for association with AMD, and the p value of 0.05 was corrected for multiple testing by this number, $p < 0.05/103{,}611 = 4.8 \times 10^{-7}$. Two SNPs, rs380390 and rs10272438, were found to be significantly associated with AMD, with a Bonferroni-corrected $p = 0.0043$ and $p = 0.008$, respectively. Upon reexamination of the quality of data and re-genotyping, rs10272438 was dismissed.

The authors went back to consider the next best hit with the third-lowest Bonferroni-corrected p value at 0.14 (>0.05), the SNP rs1329428 at 1.8 kbp (kilo base pairs) away toward the telomere from rs380390. The homozygotes for the "C" allele (major) of rs380390 and the "C" allele (major) of rs1329428 showed an increased risk of developing AMD. The two AMD-associated SNPs are located in an intron of the complement factor H (CFH) gene, on chromosome 1q31, at chr1:196731921 (GRCh38.p14) and chr1:196733680 (GRCh38.p14), respectively. Haplotype inquiries of six SNPs over a 41-kbp region lead to the identification of a disease allele conferring an increased AMD risk by a factor of 7.4 (95% CI: 2.9–19). This risk allele contains a Tyr/His change at amino acid 402 of CFH, in the SNP rs1061170 (chr1:196690107, GRCh38.p14), in exon 9 of the CFH gene. This amino acid change could potentially affect the CFH function such as binding to heparin and the C-reactive protein.

8.6.3 THE SECOND "FIRST" TRUE GWAS

The second "first" true GWAS in the world was a massive paper published in 2007 by The Wellcome Trust Case Control Consortium. In this study conducted in the British population for genetic association to seven major diseases, ~2,000 individuals for each disease and a shared set of ~3,000 controls were genotyped using the Affymetrix GeneChip 500K Mapping Array Set. Overall, 24 independent association signals under a threshold p value of 5×10^{-7} were revealed, including one for bipolar disorder, one for coronary artery disease, nine for Crohn's disease, three for rheumatoid arthritis, seven for type-I diabetes, and three for type-II diabetes. The official clock of the starting point of GWAS in the world was reset to 2007 in some recent reviews because this was the first study using a really large number of SNPs to be considered as "genome-wide," at 500,568, with 469,557 SNPs (93.8%) passing the quality control.

The authors of this paper also debated and set for almost all later GWAS the standard for multiple testing corrections. To retain the original flavor, the argument presented in the paper is quoted here:

> *Classical multiple testing theory in statistics is concerned with the problem of "multiple tests" of a single "global" null hypothesis. This, we would argue, is a problem far removed from that which faces us in genome-wide association studies, where we face the problem of testing "multiple hypotheses" (for a particular disease, one hypothesis for each SNP, or region of correlated SNPs, in the genome) and we thus do not subscribe to the view that one should correct significance levels for the number of tests performed to obtain "genome-wide significance levels."*

Instead, the prior odds of 100,000:1 to expect 10 signals in 1,000,000 "independent" regions of the genome were used for the correction, giving the threshold p value of $0.05/100{,}000 = 5 \times 10^{-7}$. The prior odds can be set by a factor of 10 in either direction depending on the expectations and levels of comfort at accepting false positive signals. In today's GWAS, the customary prior odds are 1,000,000:1, with the p value threshold of 5×10^{-8}, shown as $-\log_{10}$ (**Figure 8-7**). The variants with higher frequencies in cases than in the control group are called "hits," and the graph over the entire genome length is called a "Manhattan plot," suggestive of the skylines in Manhattan.

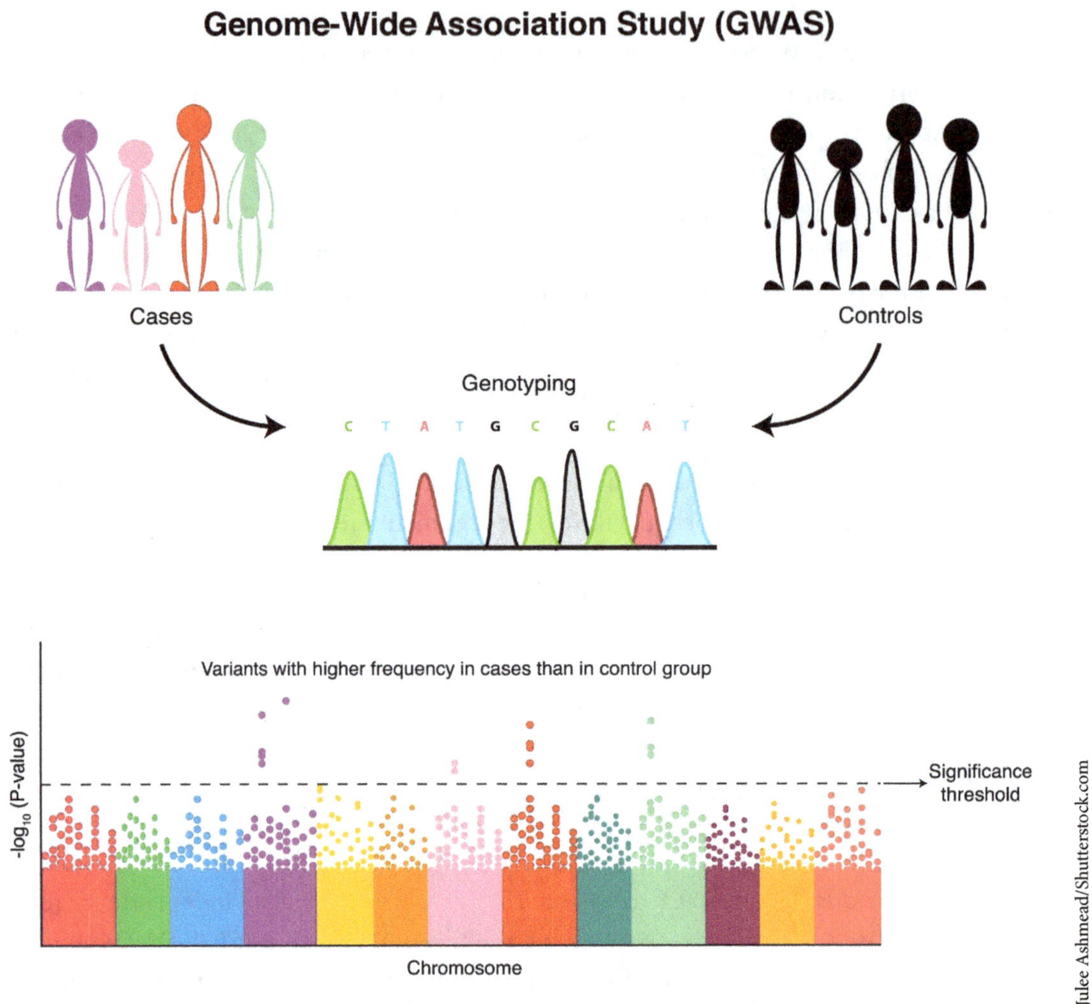

FIGURE 8-7. Genome-wide association studies

8.6.4 AN ASSOCIATION STUDY DOES NOT DETECT GENETIC ASSOCIATION

In truth, the genetic principle behind an association study, candidate gene or genome-wide, is to detect linkage disequilibrium (LD). At the practical and operational level, it is the detection of a statistically significant biased (different) combination/distribution of the marker locus alleles in the cases *versus* the controls. As we have discussed earlier, linkage disequilibrium does not require linkage, it is merely a snapshot in time of a biased allele combination/distribution across the marker locus and the unknown disease locus, different from what is expected at equilibrium, that is, random and unbiased. Unlinked loci on different chromosomes, but unbeknownst to us that they are, can present linkage disequilibrium, manifested simply as the biased allele combination/distribution across the loci.

A disease, or a subset of a complex disease caused by one genetic element, can only be truly "associated" (linked) to a single genetic locus within a linkage group, up to 50 Mbp (million base pairs) in length. A multigenic disease simply means more than one locus can produce the same disease phenotype. But all markers on all chromosomes could be "associated" (linked) before the test. The null hypothesis in a GWAS must be stated as: none of the markers to be tested is "associated" (linked) with the disease phenotype. All markers must be tested for every test subject. Thus, it is clearly not testing "one hypothesis for each SNP," but all SNPs for the same disease.

Two things are hopefully clearer at this moment. One, correcting for multiple testing by 1×10^6 will produce a large number of false positives in a GWAS, as the genetic variants are far greater than this number. Two, one must refrain from making any functional interpretations directly on the "risk alleles" "associated" with a disease phenotype.

8.7 Association Studies Are Going Back to Phenotype Genetics

As of the May 5, 2024 update, during the drafting of this chapter, the GWAS Catalog hosted by the European Bioinformatics Institute (EBI) of the European Molecular Biology Laboratory (EMBL) in collaboration with the National Human Genome Research Institute (NHGRI, USA) listed 6,850 publications on 613,731 "top associations" with 328,303 SNPs.

The field has benefitted over the past two decades from a decreasing cost and the increasing efficiency at performing the genotyping of a large number of individuals. In the 2002 "first" GWAS by Ozaki *et al.*, PCR and DNA sequencing of the PCR products were used for genotyping. By today's standards, this was labor and cost intensive and inefficient at handling a large number of samples. The 2005 "first" GWAS already moved up to a DNA microarray technology, employing the Affymetrix GeneChip Mapping 100K Set. The 2007 "first" GWAS upgraded to the Affymetrix GeneChip 500K Mapping Array Set. Today, the number of variants available for testing is about 10 million. The development of algorithms and software has also improved by orders of magnitude to aid in subject recruitment, data quality control, and data analysis.

To say that much remains to be done with the existing volume of data would be a huge understatement. But we are so emboldened to fully explore the power of "direct genetics" with new variations on the theme, such that TWAS, PWAS, and EWAS have appeared over the horizon and into research, as well as preclinical investigations. TWAS stands for transcriptome-wide association studies and uses quantitative genotype information derived from expressed RNA sequences determined by next-gen sequencing technology. PWAS stands for proteome-wide association study and quantifies the extents to which proteins are damaged by the changes resulting from genetic variants. EWAS stands for epigenome-wide association studies for the association between a phenotype and epigenetic variants, for example, changes in DNA methylation or chromatin structure/accessibility. We can view these as performing simultaneous studies along the path of the central dogma, at the genetic (DNA), gene transcription (RNA), and protein function levels, using the results from each approach to validate the findings from the others toward determining the cause of a disease. The explorations in the new WASs have only begun and the art is still being perfected.

Interestingly, and more importantly, unlike the lifeless (genomic) DNA, the RNA transcripts and the protein products encoded by the genome have biochemical functions. The mRNA does not have a direct functional output, but the expression level differences between a disease state and the normal represent real differences. Thus, the test is now "Disease/Normal" *versus* "High/Low expression," or *versus* "Yes/No function" in TWAS or PWAS, looking at the manifested changes over the genetic variants, and would yield direct association relationships with the disease phenotype. This is in principle no different from what was done in The Fly Room more than a century ago, examining the relationships between two sets of manifested traits. The key difference in human genetics is that the more than one century of time brought us the technical and technological advances and enabled the examination of phenotypes at the molecular levels, requiring the smallest amounts of patient samples. Multiple testing corrections are still required as there are thousands of genes, for example, ~20,000 code for proteins, and these studies are still prone to producing large numbers of false positives.

BIBLIOGRAPHY

1. https://www.cdc.gov/diabetes/php/data-research/index.html#:~:text=Prevalence%20of%20both%20diagnosed%20and,Table%201a%3B%20Table%201b
2. https://idf.org/about-diabetes/diabetes-facts-figures/
3. https://www.ncbi.nlm.nih.gov/snp/rs334
4. Abdellaoui A, Yengo L, Verweij KJH, Visscher PM. 15 years of GWAS discovery: Realizing the promise. *American Journal of Human Genetics.* 2023 Feb 2;110(2):179–94.
5. Armstrong RA. When to use the Bonferroni correction. *Ophthalmic and Physiological Optics.* 2014 Sep;34(5):502–8.
6. Arnheim N, Calabrese P. Understanding what determines the frequency and pattern of human germline mutations. *Nature Reviews Genetics.* 2009 Jul;10(7):478–88.
7. Ashley-Koch A, Yang Q, Olney RS. Sickle hemoglobin (HbS) allele and sickle cell disease: A HuGE review. *American Journal of Epidemiology.* 2000 May 1;151(9):839–45.
8. Bekassy Z, Lopatko Fagerström I, Bader M, Karpman D. Crosstalk between the renin-angiotensin, complement and kallikrein-kinin systems in inflammation. *Nature Reviews Immunology.* 2022 Jul;22(7):411–28.
9. Benjamini Y, Hochberg Y. Controlling the false discovery rate: A practical and powerful approach to multiple testing. *Journal of the Royal Statistical Society. Series B (Methodological).* 1995;57(1):289–300.
10. Bhupatiraju C, Patkar S, Pandharpurkar D, Joshi S, Tirunilai P. Association and interaction of -58C>T and ±9 bp polymorphisms of BDKRB2 gene causing susceptibility to essential hypertension. Clinical and Experimental Hypertension. 2012;34(3):230–5.
11. Brandes N, Linial N, Linial M. PWAS: Proteome-wide association study-linking genes and phenotypes by functional variation in proteins. *Genome Biology.* 2020 Jul 14;21(1):173.
12. Campagna MP, Xavier A, Lechner-Scott J, Maltby V, Scott RJ, Butzkueven H, Jokubaitis VG, Lea RA. Epigenome-wide association studies: Current knowledge, strategies and recommendations. *Clinical Epigenetics.* 2021 Dec 4;13(1):214.
13. Chai KX, Ward DC, Chao J, Chao L. Molecular cloning, sequence analysis, and chromosomal localization of the human protease inhibitor 4 (kallistatin) gene (PI4). *Genomics.* 1994 Sep 15;23(2):370–8.
14. David S. A current guide to candidate gene association studies. *Trends in Genetics.* 2021 Dec;37(12):1056–9.
15. Devuyst O, Bochud M, Olinger E. UMOD and the architecture of kidney disease. *Pflugers Arch: European Journal of Physiology.* 2022 Aug;474(8):771–81.
16. Fan HC, Ho LI, Chi CS, Chen SJ, Peng GS, Chan TM, Lin SZ, Harn HJ. Polyglutamine (PolyQ) diseases: Genetics to treatments. *Cell Transplantation.* 2014;23(4-5):441–58.

17. Fritz H, Jochum M, Müller-Esterl W. Kinins 1925–2000. *Biological Chemistry.* 2001 Jan;382(1):3–4.
18. Gibson G. Population genetics and GWAS: A primer. *PLoS Biology.* 2018 Mar 16;16(3):e2005485.
19. Gu D, Zhao Q, Kelly TN, Hixson JE, et al. The role of the Kallikrein-Kinin system genes in the salt sensitivity of blood pressure: The GenSalt Study. *American Journal of Epidemiology.* 2012 Oct 1;176 Suppl 7(Suppl 7):S72–80.
20. Hardy GH. Mendelian proportions in a mixed population. *Science.* 1908 Jul 10;28(706):49–50.
21. Keightley PD. Rates and fitness consequences of new mutations in humans. *Genetics.* 2012 Feb;190(2):295–304.
22. Kido T, Sikora-Wohlfeld W, Kawashima M, Kikuchi S, et al. Are minor alleles more likely to be risk alleles? *BMC Medical Genomics.* 2018 Jan 19;11(1):3.
23. Kimura A, Takahashi M, Choi BY, Bae SW, et al. Lack of association between LTA and LGALS2 polymorphisms and myocardial infarction in Japanese and Korean populations. *Tissue Antigens.* 2007 Mar;69(3):265–9.
24. Klein RJ, Zeiss C, Chew EY, Tsai JY, et al. Complement factor H polymorphism in age-related macular degeneration. *Science.* 2005 Apr 15;308(5720):385–9.
25. Li N, Liu R, Zhai H, Li L, Yin Y, Zhang J, Xia Y. Polymorphisms of the LTA gene may contribute to the risk of myocardial infarction: A meta-analysis. *PLoS One.* 2014 Mar 18;9(3):e92272.
26. Loos RJF. 15 years of genome-wide association studies and no signs of slowing down. *Nature Communication.* 2020 Nov 19;11(1):5900.
27. Maaliki D, Itani MM, Itani HA. Pathophysiology and genetics of salt-sensitive hypertension. *Frontiers in Physiology.* 2022 Sep 13;13:1001434.
28. Manolio TA, Collins FS, Cox NJ, Goldstein DB, et al. Finding the missing heritability of complex diseases. *Nature.* 2009 Oct 8;461(7265):747–53.
29. Manosroi W, Williams GH. Genetics of human primary hypertension: Focus on hormonal mechanisms. *Endocrine Reviews.* 2019 Jun 1;40(3):825–56.
30. Miyagawa T, Nishida N, Ohashi J, Kimura R, et al. Appropriate data cleaning methods for genome-wide association study. *Journal of Human Genetics.* 2008;53(10):886–93.
31. Ozaki K, Ohnishi Y, Iida A, Sekine A, et al. Functional SNPs in the lymphotoxin-alpha gene that are associated with susceptibility to myocardial infarction. *Nature Genetics.* 2002 Dec;32(4):650–4.
32. Phillips C, Amigo J, Tillmar AO, Peck MA, et al. A compilation of tri-allelic SNPs from 1000 Genomes and use of the most polymorphic loci for a large-scale human identification panel. *Forensic Science International: Genetics.* 2020 May;46:102232.
33. Salguero MV, Arosemena M, Pollin T, et al. Monogenic forms of diabetes. 2023 Dec 20. In: Lawrence JM, Casagrande SS, Herman WH, et al., editors. *Diabetes in America* [Internet]. Bethesda, MD: National Institute of Diabetes and Digestive and Kidney Diseases (NIDDK); 2023. Available from: https://www.ncbi.nlm.nih.gov/books/NBK597414/

34. Samuels ME, Friedman JM. Genetic mosaics and the germ line lineage. *Genes* (Basel). 2015 Apr 17;6(2):216–37.
35. Slatkin M. Linkage disequilibrium—Understanding the evolutionary past and mapping the medical future. *Nature Reviews Genetics.* 2008 Jun;9(6):477–85.
36. Schriml LM, Lichenstein R, Bisordi K, Bearer C, Baron JA, Greene C. Modeling the enigma of complex disease etiology. *Journal of Translational Medicine.* 2023 Feb 25;21(1):148.
37. Suzuki K, Hatzikotoulas K, Southam L, Taylor HJ, et al. Genetic drivers of heterogeneity in type 2 diabetes pathophysiology. *Nature.* 2024 Mar;627(8003):347–57.
38. Uffelmann E, Huang QQ, Munung NS, et al. Genome-wide association studies. *Nature Reviews Methods Primers.* 2021;1:59.
39. Wellcome Trust Case Control Consortium. Genome-wide association study of 14,000 cases of seven common diseases and 3,000 shared controls. *Nature.* 2007 Jun 7;447(7145):661–78.
40. Yang Y, Chan L. Monogenic diabetes: What it teaches us on the common forms of type 1 and type 2 diabetes. *Endocrine Reviews.* 2016 Jun;37(3):190–222.

9 Genetics in Modern Medicine

9.1 Gene Therapies

9.1.1 CLONING THE LARGEST HUMAN GENE, DYSTROPHIN

Just as I was drafting this chapter, this headline on May 8, 2024, from my daily email alert in Genetic Engineering & Biotechnology News (GEN) caught my attention: "Boy Dosed with Pfizer DMD Gene Therapy Dies A Year after Phase II Trial." DMD is Duchenne muscular dystrophy, an X-linked recessive disease manifested only in boys, with one copy of the X chromosome. DMD was named after the French neurologist Guillaume Benjamin Amand Duchenne (1806–1875) who documented the case of a boy with this condition in 1861, described by him as "congenital hypertrophic paraplegia." Duchenne provided a more accurate description in 1868 and named it "pseudohypertrophic muscular dystrophy."

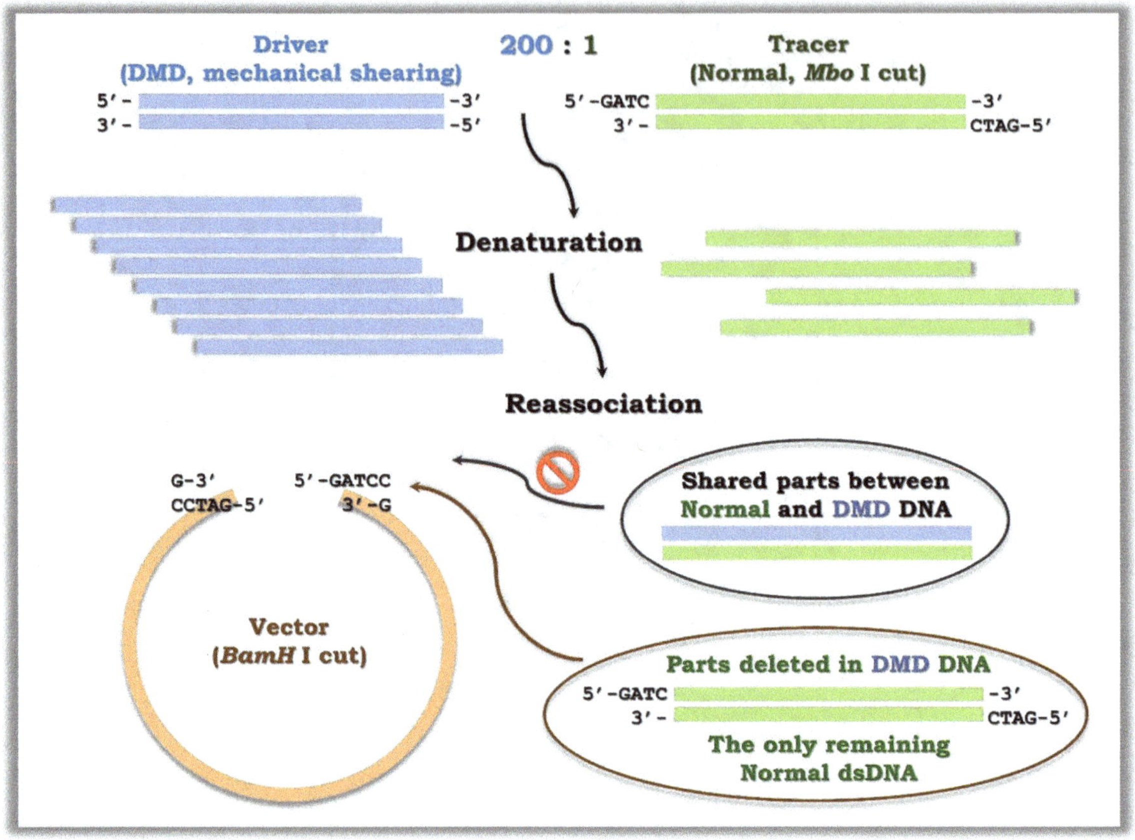

FIGURE 9-1. Subtractive cloning of the DMD (dystrophin) gene (fragments)

The cloning of the DMD (dystrophin) gene began with a masterpiece of ingenuity by "subtracting" the DNA from a "normal" source using the DNA from a DMD patient, a boy whose X chromosome had a small deletion in the short arm. The work was published in 1985 by Louis Kunkel's lab of Harvard Medical School. The "normal" source was a human lymphoid cell line harboring three extra X chromosomes (49,XXXXY). The DNA from the cell line, called the "tracer DNA," and the patient, called the "driver DNA," were fragmented by different methods to generate "sticky ends" by restriction digestion (*Mbo* I) or "blunt ends" by mechanical shearing, respectively (**Figure 9-1**). The blue driver DNA at a 200-fold quantitative excess and the green tracer DNA were both strand-separated and mixed. At such an overwhelming quantitative ratio, the reassociation process would ensure that all DNA fragments in the green tracer are bound away by the blue driver, with the hybrid double-stranded DNA having the hybrid form of ends. The 199-times redundant blue driver DNA reassociated among themselves to blunt-ended double-stranded DNA. The only parts of the green tracer DNA that could reassociate among themselves were the fragments from the parts deleted in the blue driver, the patient's DNA, that is, no matching parts in the blue DNA to bind away the green DNA. This was the portion of the normal DNA left over after the "subtraction," and the only double-stranded DNA with restored "sticky" restriction enzyme ends. And they were the only DNA that could be cloned into a vector DNA modified by a restriction enzyme (*BamH* I) with compatible ends.

One such tracer DNA fragment hybridized to a large mRNA transcript in the RNA of human fetal skeletal muscle cells and enabled the isolation of candidate cDNA clones of the DMD gene transcripts by the Kunkel lab in 1986. One year later (1987), the Kunkel lab revealed the complete genomic and genetic landscape at the DMD locus. Today, after years of refinement, we have the full-length DMD gene, named dystrophin, with 79 exons at ~2.4 Mbp (million base pairs), the largest known human gene. The time required for transcribing and co-transcriptional splicing of the dystrophin mRNA is 16 hours. The mature mRNA transcript is ~14 kb, encoding the 427-kDa (kilodalton) full-length dystrophin protein. Dystrophin links the cytoskeleton to the extracellular matrix (ECM) to transmit the force from the contractile elements inside the cell to the ECM structures while maintaining the muscle fiber integrity, as a molecular shock absorber. An absence of a functional dystrophin in skeletal or cardiac muscle cells will compromise membrane integrity and lead to muscle fiber damage during contraction, and further to progressive muscle wasting and loss.

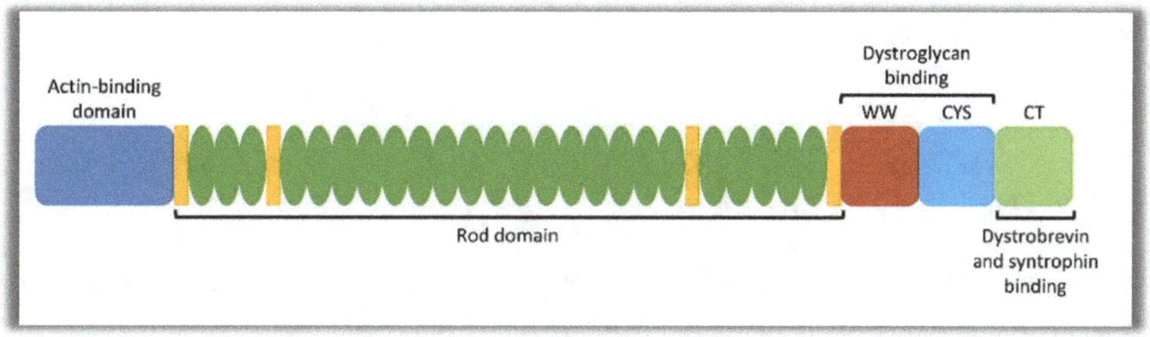

FIGURE 9-2. The functional domains of dystrophin

9.1.2 BIGGER IS NOT BETTER—THE GENETIC DEFECTS IN THE HUMAN DYSTROPHIN GENE

The dystrophin protein has two essential parts on the two ends of its polypeptide chain, the actin-binding domain on the amino terminus and the dystroglycan-binding and syntrophin- and dystrobrevin-binding domains on the carboxyl terminus (**Figure 9-2**). In the middle are 24 rod domains that function as the spring in a shock absorber for an automobile, and there is a good degree of partial redundancy in the rod domains. But reading frame interruptions in the long dystrophin coding region will inevitably terminate the translation before the carboxyl-terminal essential domains, producing truncated forms of dystrophin with no function at all.

The massive gene is laden with several thousand reported mutations, about one-third *de novo* and the other two-thirds familial. The mutations range from deletions (60%–70%), duplications (5%–20%), point mutations and small insertions or deletions (5%–20%), and rare translocations with other chromosomes, depending on the population and disease type. Some of these mutations cause the Becker muscular dystrophy (BMD) with similar symptoms to DMD, but with a slower and less severe disease progression. Critically, these BMD mutations do not alter the transcript reading frame but produce shorter and partially functional dystrophin forms. The analogy would be to have at least a shorter shock absorber, instead of having none. In extreme cases, BMD patients could lose up to 46% of the codons by deletion and still manage a limited functionality with the much shorter variant dystrophin protein.

9.1.3 SHORTER IS BETTER THAN NOTHING—THE STRATEGIES OF DMD GENE THERAPY

A strong correlation between the disease phenotype and whether a deletion mutation alters the reading frame was revealed by analyzing several hundred mutations at the DMD/BMD locus. Many "in-frame" deletions of the DMD gene can remain undetected as the individuals would appear more or less asymptomatic or without the characteristic DMD/BMD clinical features. The "reading frame hypothesis" was inspirational for gene therapy research aimed at manipulating and engineering to repair the defective DMD genes in affected patients. The genetic engineering design behind these gene therapy efforts is fairly simple and straightforward: reading frame repair to produce a functional or partially functional dystrophin protein. The part of the dystrophin protein and the corresponding coding regions in the gene that is suitable for such a repair strategy is in the rod domains. By cutting and splicing, the out-of-frame parts of the coding region in the gene are removed and the ends are reconnected in-frame to make a shorter dystrophin but with the essential domains on both ends. Up to 60%–80% of the DMD mutations could be corrected by "gene editing" using technologies such as the CRISPR/Cas9 system (**Figure 9-3**) and the newer variations on the theme. Initial successes were achieved in mouse DMD models harboring specific mutations found in human patients, and these were followed by the success in a naturally occurring larger animal DMD model in the dog. The gene-editing molecular reagents are carried by adeno-associated viruses (AAV) as a vehicle, and delivered by intravenous infusion. The viral vectors carry regulatory elements for muscle-specific expression.

Gene editing takes away the need to completely replace a defective DMD gene, but it will not work for the mutations that cannot be repaired by gene editing, such as large deletions that result in the loss of the essential domains. For the DMD gene with its massive size of 2.4 Mbp (million base pairs), a total replacement strategy is impossible as this is way larger than the capacity of any viral vectors employed for gene therapy. The dystrophin cDNA, a double-stranded DNA copy of its mRNA at ~1.4 kbp (kilo base pairs), is still too big. The lessons learned about the rod domains from the BMD patients inspired the designs of abbreviated or shorter dystrophin genes for DMD gene therapy, the mini-dystrophin gene (cDNA) at 6–8 kbp and the micro-dystrophin gene (cDNA) at 3.5–4 kbp. The micro-dystrophin gene can be accommodated by the AAV vector, which has a capacity of 5 kbp but the mini-dystrophin gene requires a two-part split for a dual-AAV delivery. The main drawback of the abbreviated or shorter genes is the limited functionality of the shorter dystrophin proteins produced. Whereas for the gene editing approach, issues such as off-target hits, that is, cutting and splicing at unintended sites in the genome, and immunogenicity against the bacterial proteins of the CRISPR/Cas9 system need to be resolved before clinical applications.

Behind the GEN news alert headline was a boy (age not disclosed) in the Phase II DAYLIGHT trial NCT05429372 on the investigational gene therapy fordadistrogene movaparvovec (PF-06939926), an AAV9 vector carrying a human mini-dystrophin gene under the control of a muscle-specific promoter. This was the second death associated with fordadistrogene movaparvovec during a clinical study, with the first reported in 2021.

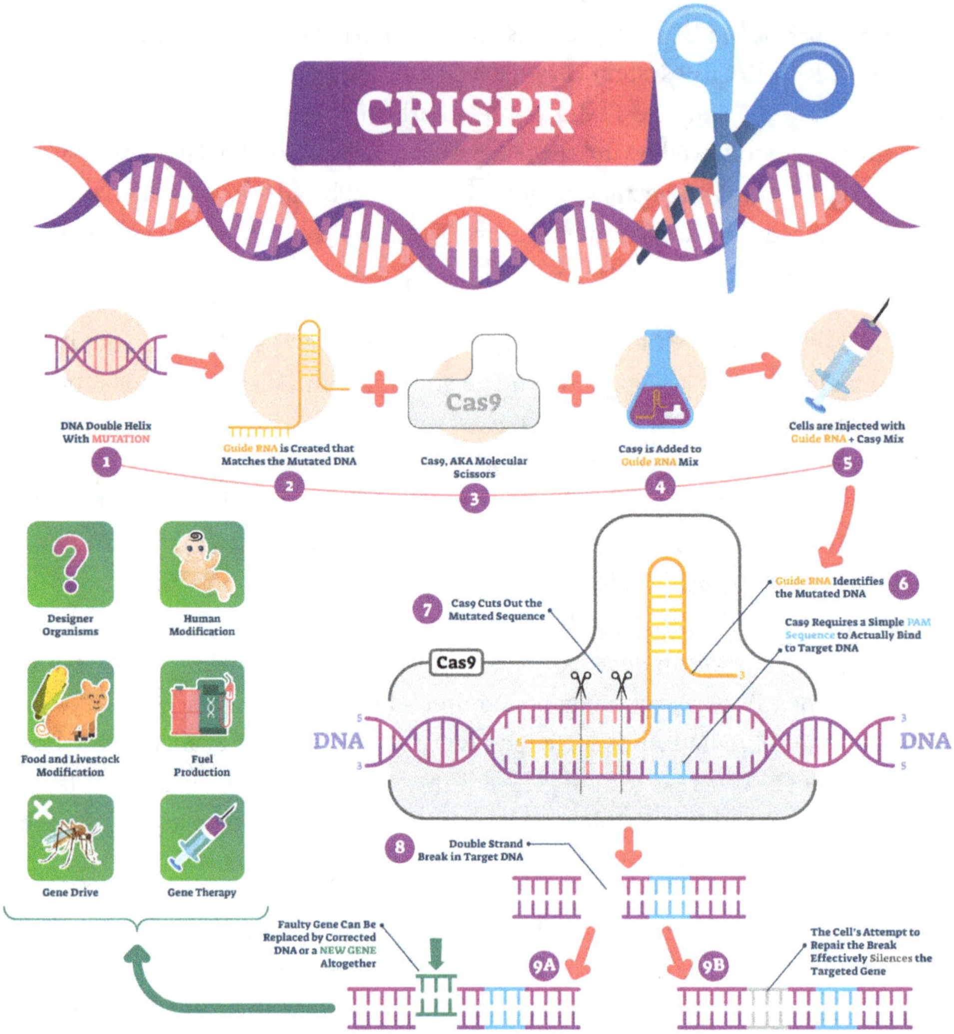

VectorMine/Shutterstock.com

FIGURE 9-3. Gene editing via the CRISPR/Cas9 system

9.1.4 HOPES AND PROMISES OF HUMAN GENE THERAPY

Human gene therapy has come a long way since its early approvals and trials in the late 1980s and the early 1990s. The idea has evolved over more than four decades of time in development and application. At its core, it involves introducing a nucleic acid as a therapeutic agent into a cell to remedy an inborn genetic defect or a gained phenotype from somatic mutations, such as those in sporadic cancers. Currently, there are 16 different forms of nucleic acid agents used for gene therapy, including siRNA, miRNA, antisense oligonucleotides, components of the CRISPR/Cas9 system, plasmid DNA carrying a specific gene or cDNA, mRNA, or genes encoding various DNA modifying enzymes. A variety of 46 different vector or delivery systems are employed, including retroviral and AAV vectors, PEGylated nanoparticles, and lipid nanoparticles (LNPs). Approved human gene therapy products target a wide ranges of diseases, including many forms of cancer, and hereditary disorders, such as β-thalassemia, hemophilia B (congenital factor IX deficiency), aromatic L-amino acid decarboxylase (AADC) deficiency, and ADA-SCID (severe combined immunodeficiency due to adenosine deaminase deficiency). The successful treatment of ADA-SCID in a four-year-old girl, Ashanthi DeSilva, in 1990 by W. French Anderson and colleagues was celebrated as the first in human gene therapy.

Unlike conventional therapies with a drug, which would still require a very stringent criteria for use, gene therapy may need to be contemplated on a patient by patient basis due to the special issues unique to this form of intervention. On the technical end, issues such as safety, specificity, delivery efficiency, and immune responses remain as major challenges, as we can see from the brief introduction of the current states in DMD gene therapy. On the economic end, the cost of gene therapy can be very high. For example, the one-time curative Zolgensma (onasemnogene abeparvovec-xioi) approved by the U.S. FDA in 2019, an AAV9-based human SMN1 (survival of motor neuron 1) gene therapy for infants with spinal muscular atrophy (SMA), has a cost of ~$2 million!

9.2 Pharmacogenetics, Pharmacogenomics, Personalized Medicine, Precision Medicine

9.2.1 PHARMACOGENETICS

This portmanteau has two words, pharmacology and genetics, as in "the study of drug effects" in relation to "DNA variations." The intent is twofold: one, to improve the efficacy of drugs administered to patients, and two, to reduce or eliminate adverse drug effects (ADRs). Estimates of ADRs place them behind heart disease, cancer, and stroke as the fourth most common cause of death worldwide. The first recognized genetic cause for an ADR was the deficiency of glucose-6-phosphate dehydrogenase (G6PD), an X-linked recessive disorder. The ADR condition is known as "favism" when the affected individuals experience potentially fatal hemolytic anemia after consuming fava beans, and can be traced back to 510 BC in the records of Pythagoras. G6PD is a so-called housekeeping enzyme in the basic metabolism of all cells in the body and it plays a vital role to protect cells from oxidative stress damages induced by reactive oxygen species (ROS). The red blood cells

(RBCs) are highly vulnerable to oxidative stress damages as they are responsible for oxygen transport. Deficiencies of G6PD can result in acute hemolytic anemia with an increased ROS production, for example, by the consumption of foods such as fava beans or administration of certain drugs that contain high amounts of oxidizing chemicals (**Figure 9-4**). There are an estimated 400 million people affected by G6PD deficiency worldwide, making it the most frequent genetic deficiency in humans. More than 40 foods and drugs are to be avoided by this population, including drugs as common as acetylsalicylic acid (Aspirin) and acetaminophen (Tylenol).

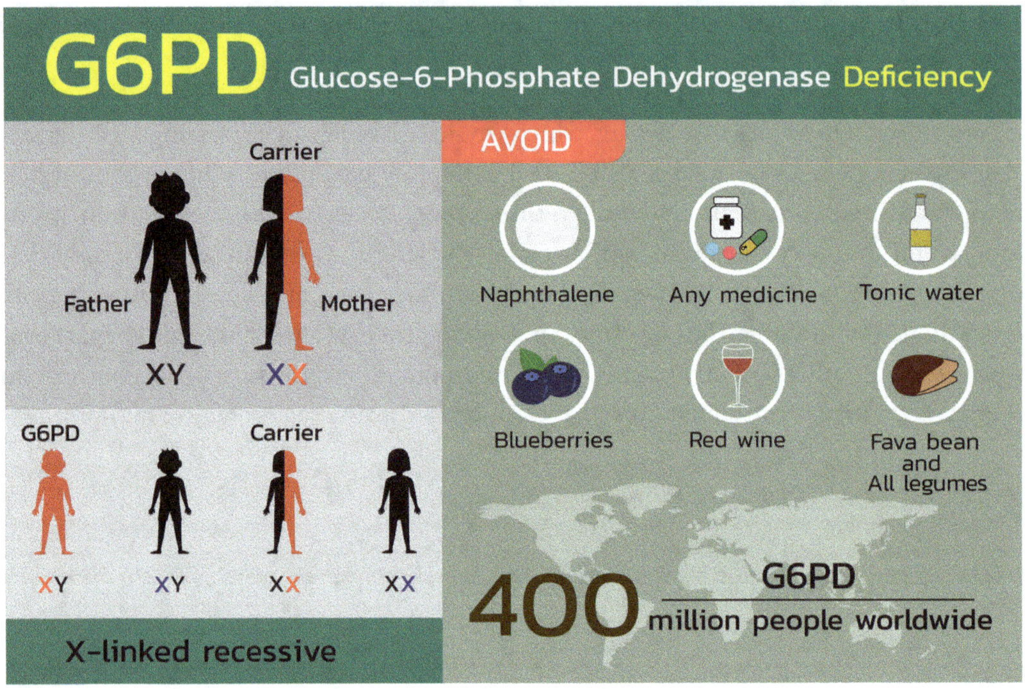

KOHYAO/Shutterstock.com

FIGURE 9-4. Glucose-6-phosphate dehydrogenase (G6PD) deficiency

9.2.2 PHARMACOGENOMICS

This is a newer and trendier but almost interchangeable term with pharmacogenetics, emphasizing on more of a genome-wide view of the genetic factors that affect drug efficacy or induce ADRs. It has gained a flavor as using the next-gen genome science technologies for the discovery and development of new drugs, or the optimization of existing drugs to maximize efficacy and minimize toxicity.

The autosomal recessive disorder cystic fibrosis (CF) is caused by mutations in the *CFTR* gene, with more than 2,000 reported, causing varying clinical consequences. The most common mutation is the ΔF508, a deletion of the Phe/F residue in amino acid position 508, found in 70% of the CF cases. Over the years since the success of cloning the *CFTR* gene via reverse genetics, multidisciplinary approaches have been taken to treat or help alleviate the symptoms of the patients. These include efforts in gene therapy to replace or edit the faulty CFTR alleles and in regulating other ion channels to counteract the defective chloride transport, for example, via suppressing the epithelial sodium channel (ENaC) or its activating proteases, for example, prostasin/CAP1/PRSS8. A new class of small molecule drugs known as CFTR modulators has been the most successful in the clinical setting from among all of the approaches. These CFTR modulators were discovered by high-throughput screening using *in vitro* systems assembled to manifest the specific genetic defects in the CFTR protein function and identify compounds that correct the defects or improve the function. The CFTR modulators can be mutation-specific and grouped according to their mechanisms of action, including potentiators, correctors, stabilizers, and amplifiers (**Figure 9-5**).

The Gly551Asp missense CFTR mutation is a "gating" mutation, with a defective channel opening action of the cell-surface CFTR protein. Ivacaftor is a drug of the potentiator group which keeps the CFTR chloride channel in an "open" state, with the U.S. FDA approval to target the Gly551Asp and several other "gating" mutations. The ΔF508 deletion causes defects in protein folding and leads to degradation, and requires the actions of correctors. These include lumacaftor, tezacaftor, and elexacaftor, which target abnormal protein folding to improve channel presentation at the cell surface. The correctors can be used in combination with the potentiators, for example, an elexacaftor/tezacaftor/ivacaftor combination. Stabilizer modulators such as cavosonstat also target the ΔF508 mutation and can be used together with potentiators and correctors. The amplifier modulators, for example, nesolicafto, are not mutation-specific but they increase the mRNA stability to enhance the CFTR protein synthesis. Again, these can also be used in combination with any or all other types of modulators. Ivacaftor produced a 55% reduction of the pulmonary exacerbation rate in patients carrying the Gly551Asp mutation. The combination of elexacaftor/tezacaftor/ivacaftor could achieve a 63% exacerbation rate reduction in patients carrying at least one ΔF508 allele, presenting a haploinsufficiency phenotype.

The true potential intended in the change of the term from pharmacogenetics to pharmacogenomics will be in the application of genome-wide association studies (GWAS) for the discovery of gene–drug interactions, using drug response as the phenotype.

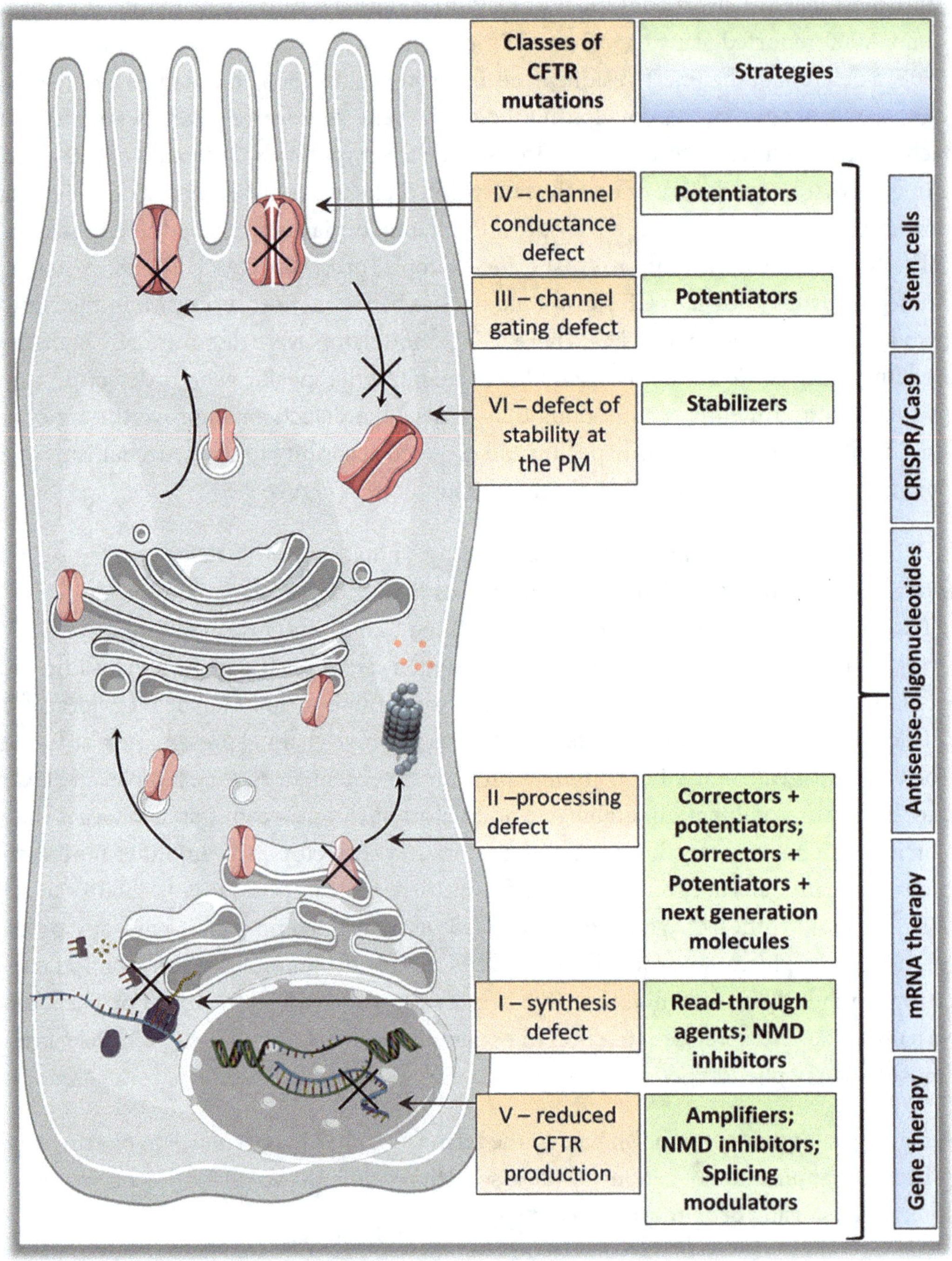

FIGURE 9-5. CFTR modulators can address mutation-specific defects

9.2.3 PRECISION MEDICINE/PERSONALIZED MEDICINE

Precision medicine is the newer term for personalized medicine (**Figure 9-6**), which "is an innovative approach (to medicine) that takes into account individual differences in patients' genes, environments, and lifestyles" (from "NIH...Turning Discovery Into Health," National Institutes of Health). This approach is in direct contrast to the traditional "consensus" medicine based on the average responses to treatment regimens and drugs. Most advances have been made in the areas of "precision oncology" and "cancer immunotherapies," and the study of "rare diseases." A rare disease affects less than 1 in 2,000 live births but there are between 7,000 and 8,000 rare diseases, affecting an overall large population at 25–30 million in the United States.

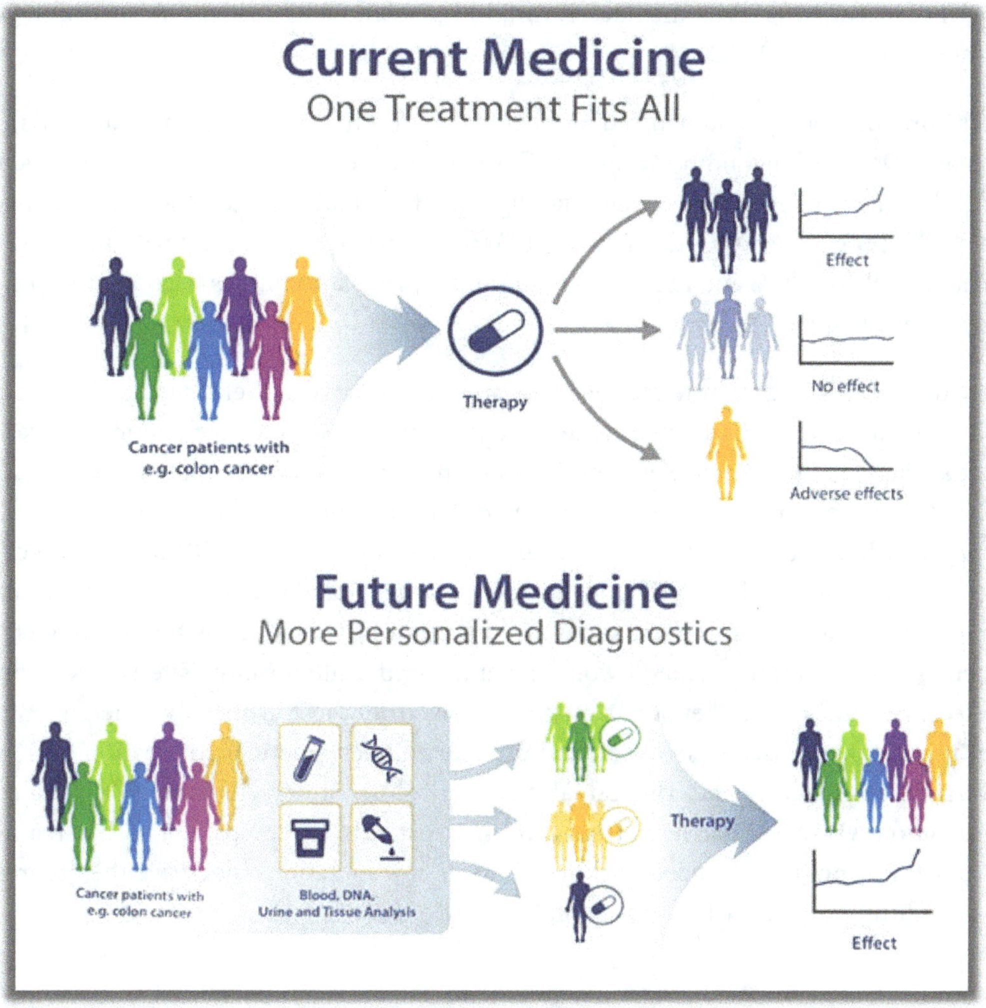

FIGURE 9-6. Conventional "consensus" medicine *versus* precision "personalized" medicine

Credit: *Sofia Jones*: https://pressbooks.pub/anne1/chapter/personalized-medicine/

Early successes in precision oncology include targeting the BCR-ABL rearrangement in chronic myeloid leukemia (CML) by the small molecule drug imatinib (Gleevec or Glivec), a tyrosine kinase inhibitor (TKI), approved by the U.S. FDA in 2001. More than 95% of the CML cases contain an abnormal chromosome from a reciprocal translocation between chromosome 9 at band q34 and chromosome 22 at band q11, known as the Philadelphia (Ph) chromosome for the city of its initial discovery. The translocation creates a fusion of the breakpoint cluster region (BCR) gene with the ABL tyrosine-protein kinase gene, the BCR-ABL oncogene. Gleevac achieved an overall survival of 89% at five years for CML patients. This class of gene or mutation-specific drugs have seen a big growth in application, with the best examples in the treatment of non-small cell lung cancers (NSCLCs) carrying specific mutations of the epidermal growth factor receptor (EGFR). The globally approved EGFR-targeting TKIs include erlotinib (Tarceva), gefitinib (Iressa), afatinib (Gilotrif), and osimertinib (Tagrisso).

In cancer immunotherapy, a humanized monoclonal antibody pembrolizumab (Keytruda) was the first human PD1–blocking antibody approved for use in the United States to treat patients with unresectable or metastatic melanoma after failing other therapies. It targets the immune checkpoint molecule PD1 (programmed cell death protein 1) to prevent the engagement with its ligand PD-L1, which cancer cells produce and use to evade immune surveillance. It has been approved to treat a full range of other cancers today.

Rare genetic diseases "escape" the conventional approaches for discovery, such as linkage analysis or GWAS, due to the fact that they are rare. The data generated by whole-genome or exome (exons) sequencing, transcriptomics, proteomics, and epigenomics will continue to aid the diagnosis and characterization of rare diseases. A strategy of "getting ahead of the game" is on the horizon to potentially help solve the mystery of rare genetic diseases. Using the CRISPR/Cas9 technology, genome-wide random but site-specific mutations can be generated in cultured cells or animal models. Random refers to the experimental design of generating mutations without a predetermined site, whereas any mutation generated would be at a specific site once the site is determined. It is anticipated that some of the phenotypic manifestations from these artificially generated mutations could mimic a human disease condition and provide a direct genetic identification of the disease mechanism or a candidate gene. This can also be applied to "prescreen" cancer drugs to cover for cancer mutations yet to be discovered from cancers in patients. Once a patient with a treatment-resistant cancer is genome-sequenced to show identical new mutations that drive the progression of the cancer, a drug would have been ready off the shelf.

9.3 Molecular Diagnostics

The *Journal of Molecular Diagnostics* is the official publication of the Association for Molecular Pathology (AMP), launched in November 1999, to report "scientific advances in the translation and validation of molecular discoveries in medicine into the clinical diagnostic setting, and the description and application of technological advances in the field of molecular diagnostic medicine." The two words that describe the field are about the most precise and concise: molecular diagnostics. The molecules used for the diagnosis of diseases are literally everything that is part of a human body, organic and inorganic.

The intent of the *Journal of Molecular Diagnostics* at its launching was clearly focused on the molecules of modern molecular biology, narrowly understood and accepted to mean the nucleic acids in the central dogma, DNA and RNA. The editorial comment of the inaugural issue had the following statement: "The identification of new genes and the development of new methodology for gene expression analysis, together with conceptual advances, continue to expand at a very rapid pace. The new and exciting frontier is the application of this knowledge to the prevention, diagnosis, and treatment of disease. The field of molecular diagnostics is still in its infancy but is here to stay and is gradually changing the face of medicine." "Molecular diagnostics" started with a heavy flavor of DNA testing, and even today, the first impression upon hearing or seeing this term would still be "molecular genetic testing of DNA."

9.3.1 THE EVOLUTION OF MOLECULAR DIAGNOSTICS

Diagnosis of diseases has always relied on the presence or absence, or levels of molecules in or from a human body. When a patient is checked into a doctor's office or a hospital, two bodily specimens are often collected for analysis to aid the diagnosis of the patient's condition: blood and urine. In the blood, molecules such as cholesterol (the "good" (HDL) and the "bad" (LDL)), triglycerides, lipids, and sugar (glucose) are routinely measured, in addition to the blood cell counts. A rapid urine test can give the results on the presence and levels of a range of molecules that should not be present in the urine, such as proteins, sugar (glucose), nitrite, ketone, bilirubin (a breakdown product of hemoglobin), and urobilinogen (a breakdown product of bilirubin). Their presence to the levels beyond the accepted normal thresholds indicates a disease state of the patient. In ancient times several thousand years ago, the observation of sweet-tasting urine that attracted ants could be recognized as the primitive form of molecular diagnosis of diabetes for the individuals giving the urine samples.

The brand of molecular diagnostics at the turn of this century/millennium became far more informative with the accumulation of our knowledge and understanding of the human body and its disease states, and with the advancements in biotechnology. The first molecular genetic diagnostic test can be credited to Sir Archibald Garrod in 1902, with his observation of black urine from newborn babies and reasoning of this condition, alkaptonuria, to conform to the Mendelian pattern of inheritance. We have already learned that the black urine was caused by the incomplete metabolism of the amino acids tyrosine and phenylalanine due to the absence of an enzyme and marveled at the number of years (>90) it took to map and clone the gene of the missing enzyme, homogentisic acid oxidase (HGD). Today, to aid the validation of the clinical findings, alkaptonuria is diagnosed molecularly with more sophisticated instrumentation and more accurate results, no longer dependent on the urine color change, which is not disease-specific. The amounts of homogentisic acid (HGA) in the 24-hour urine is measured by gas chromatography-mass spectrometry (GC-MS). More important, the addition of molecular genetic testing can identify the specific mutations in the *HGD* gene in the patients and family members for genetic counseling.

The very first DNA molecular genetic test for a hereditary human disease, sickle cell disease (SCD), was developed in 1983 on the platform of a Southern blot hybridization, named after Sir Edwin Southern, Professor Emeritus of Biochemistry, University of Oxford, England. The Southern blot involves a reproducible way to fragment the large human genomic DNA into smaller (<20-kbp) pieces, which are resolved by size in agarose gel electrophoresis. The resolved DNA in the gel is then strand-separated and transferred, by capillary action in Southern's original design, or by force such as pressure or vacuum, onto a solid support, such as a nitrocellulose filter. The open area on the filter without the test sample DNA is saturated with a "background DNA" to prevent the probe DNA from binding. Fish sperm DNA, first studied by the discoverer of DNA Friedrich Miescher, was the most commonly used, behind the idea that fish DNA sequences would be very different from human DNA sequences thus a probe for a human gene would not bind to the fish DNA.

The immobilized sample DNA is "hybridized" with a probe DNA (from a different source, hence "hybridized"). The filter will be stringently washed to remove nonspecific binding of the probe, leaving only the sequence-specific signals, revealed via a radioisotope incorporated in the probe DNA in an autoradiogram on an X-ray film. Short DNA probes at 19 nucleotides in length, called oligonucleotides, corresponding to the normal and sickle alleles of the β-globin gene were used to clearly and unequivocally determine the genotypes of individuals, as homozygous normal, homozygous sickle, or heterozygous. For the latter group, that is, individuals with the sickle cell trait (SCT), the genotype information is vital as their conditions are not manifested but can be quickly induced during oxygen deprivation, for example, in strenuous excises. This technique was referred to as using "allele-specific oligos" (oligonucleotides).

9.3.2 THE PRACTICAL CONSIDERATIONS OF MOLECULAR DIAGNOSTICS

Years ago in an interesting conversation with an MD-PhD pathologist friend, he said the following about methods that he wants, or does not want for his practice: "If it costs more than $5 or you cannot teach me to do it in 5 minutes, I don't want it." This was an exact quote I cannot forget because it was so unique and I have been quoting it over the years. It may be a bit excessive or even extreme, but the excessiveness helped add power to the message. The prototype molecular genetic diagnostic test for SCD by genotyping the β-globin alleles with or without the A→T (E6V) point mutation would have costed much more than $5 and my pathologist friend would have to spend no less than a week's time to get the results back.

The first step is the purification of human genomic DNA from the white blood cells (WBCs, *aka*, leukocytes). A typical endeavor at the time required a donation of 50 ml of blood, and the procedures used, known as phenol-chloroform extraction and alcohol precipitation, were not very different from those used by Friedrich Miescher in 1869 to isolate his "nuclein" (human genomic DNA) from the pus on wound bandages. The procedures required two days to complete (Day 1 to Day 2). The samples applied to the Southern blot were in the range of 10 μg (3 million copies of the genome) *per* individual, and the use of oligonucleotides as probes was always challenging for a number of technical reasons. The fragmentation of the genomic DNA using restriction endonucleases was performed overnight, followed by gel electrophoresis the next day, whole day (Day 3). The Southern transfer by capillary action would take another overnight (Day 4), and the hybridization over another night, still (Day 5). Stringency wash took the day after (Day 5), and the exposure to X-ray film at −80°C could take between two (Day 7) to five days (Day 10). I recall with a strange sense of fondness, as I write this, the many weeks and months that I spent personally to genotype rather small groups of individuals at candidate genetic loci for linkage to blood pressure regulation, using the Southern blot technique. This seemingly painful long text recalling the steps (we have not mentioned the costs) should not be viewed as a critique of the "founding fathers" of molecular genetic tests. This was the proof of principle.

9.3.3 THE INVENTION OF THE POLYMERASE CHAIN REACTION CHANGED MOLECULAR DIAGNOSTICS

A major leap forward in modern molecular biology and genetics on the technical and technological fronts was the invention of the polymerase chain reaction (PCR) by the late Kary B. Mullis (1944–2019), who received a half share of the Nobel Prize in Chemistry in 1993 for this innovation. His Nobel Prize page described the PCR and its impact like this:

In 1985, Kary Mullis invented the process known as polymerase chain reaction (PCR), in which a small amount of DNA can be copied in large quantities over a short period of time. By applying heat, the DNA molecule's two strands are separated and the DNA building blocks that have been added are bonded to each strand. With the help of the enzyme DNA polymerase, new DNA chains are formed and the process can then be repeated. PCR has been of major importance in both medical research and forensic science.

Mullis was a coauthor of a 1985 *Science* paper that described the significant improvement in the workflow of the allele-specific oligo molecular genetic testing for SCD, with the conclusion statement in the paper's abstract: "The beta-globin genotype can be determined in less than 1 day on samples containing significantly less than 1 microgram of genomic DNA." The estimated fold of DNA amplification was 220,000. This improved test still required two steps, amplification and hybridization. In 1989, the group that provided the proof of principle, led by Bruce Wallace of the Beckman Research Institute of the City of Hope, designed allele-specific primers to carry out allele-specific PCR (ASPCR) for the SCD genotyping, omitting the hybridization step.

Today, the sample requirement by the cutting-edge technology for amplification-based analysis is down to literally two copies, the genomic DNA in a single cell! These are called "single-cell genomic, epigenomic, transcriptomic, and proteomic methods," but the technical designs are still based on the same principles behind the early versions of molecular genetic testing methods. Allow me to self-quote: ". . . there are only two things we can do to DNA, *in vivo*, or *in vitro*, from the foundations of a Southern blot to beyond next-gen sequencing. One, we can separate and reassociate the DNA strands, and two, we can make short DNA long and long DNA short."

The headline in the daily email alert on May 16, 2024 from Genetic Engineering & Biotechnology News was: "Rapidly Aging Brain Cells Discovered Through Single-Cell Mapping in Alzheimer's Disease." The source paper of this news report was published in *Nature* by a research team in University of California, San Diego, titled "Single-Cell Multiplex Chromatin and RNA Interactions in Aging Human Brain." The study used a novel technique, multinucleic acid interaction mapping in single cells (MUSIC), "for concurrent profiling of multiplex chromatin interactions, gene expression, and RNA-chromatin associations within individual nuclei." In plain language, it allows the capture of freeze-frame photos on the genetic material itself, the DNA, in the real-time gene transcription activities in postmortem human brain tissues on a single-cell basis, also called "spatial biology" as it is done in the proper tissue context. Correlative findings suggested signature profiles of gene transcription activities in the aging brain cells, and of Alzheimer Disease, the latter was based on a sample size of 7 out of a total of 14.

More than twenty years ago, I was privileged to have met Kary Mullis in person when he was invited to our university to give a talk. I was curious about an alleged quote from him: "Don't ask me about PCR, I no longer recognize it," which was meant as a superpositive response, attributed to him, to the plethora of molecular amplification techniques inspired by the PCR. I asked him if the quote was true, and he replied with a wicked smile . . . just like that in the picture we took with him. Standing by Kary on the other side in the picture is my wife and life-long collaborator Dr. Li-Mei Chen.

The second story in the daily email alert on May 15, 2024 from Genetic Engineering & Biotechnology News was: "Could Single Technologies' 3D Sequencing Deliver the $10 Genome?" The reporter used the term "game changer in NGS technology," next-gen sequencing technology, to emphasize on the innovation, but many criticisms and doubts remain. Many obstacles need to be overcome, some technical and some financial. The rough estimate on the cost to generate a human whole genome sequence has, however, been dramatically reduced over the past two decades, from $10 million in 2007 to below $1,000 today. This is not quite a $5 test that can be done in 5 minutes, to the disappointment of my pathologist friend. But the information generated, along with data from all the other omics on the overall molecular landscapes of a patient, presents both an opportunity and a challenge in the pursuit of precision medicine.

9.4 Organ Factories

The headline in the daily email alert on March 22, 2024 from Genetic Engineering & Biotechnology News was: "First Genetically Engineered Pig Kidney Transplanted into Living Patient." The transplanted "human-compatible" genetically engineered pig kidney was developed by the biotech company eGenesis, and the surgery was performed by a team of doctors at the Massachusetts General Hospital (MGH). The candidate patient of the "xenotransplantation" was at a "terminal" stage of chronic kidney disease (CKD), caused by type-II diabetes and high blood pressure, which probably also led to the failure of a transplanted human kidney five years prior. The patient developed complications after resuming regular kidney dialysis to require frequent hospital visits for management of his health.

Kidney transplant is the most successful among all solid-organ transplants, with more than 270,000 procedures performed since 1970, and the median survival time is 12.4 years. The median survival time is 5.4 years for patients on the waiting list for a donor. According to the stats provided by the Health Resources & Services Administration, in the United States, 17 people die each day waiting for an organ transplant. As of March, 2024, 89,101 patients are on the waiting list for a kidney. The urgent need for the waiting patients and those whose transplanted organs failed inspired the explorations of animal sources for organs.

Even with human donors, organ transplant faces a tremendous challenge of rejection, a bodily immune response to the "foreign organ," acute or chronic. At the molecular genetic level, the mechanisms that initiate and maintain the immune signals to reject the transplant have been well studied. Among these, the differences in the major histocompatibility complex (MHC) genes are the key for recognizing "nonself" molecules in allogeneic (same species) and xenogeneic (different species) organ transplantation. The MHC, synonymous to the human leukocyte antigen (HLA), is encoded by the most polymorphic gene cluster in the human genome, on the short arm of chromosome 6 (6p21.3). It is called a "super locus" with more than 140 genes that can regulate the transplantation immune response. HLA tissue typing is performed to match the recipient with the organ donor for compatibility in the HLA antigens.

To make cross-species xenotransplantation molecularly compatible, the scientists at eGenesis modified the kidney donor pig genome by CRISPR/Cas9 at 69 sites to eliminate three pig genes and insert seven human transgenes ("ten-GE," for 10 engineered genes), and to inactivate 59 viral genes. The organ was tested for supporting the kidney functions and survival in monkeys, lasting up to more than two years. Another major step was taken before the genetically engineered pig kidney was transplanted in a living human recipient. In September 2021, a kidney from a genetically modified pig, "GalSafe," with a pig gene knockout (α1,3-galactosyltransferase) was transplanted into a brain-dead recipient by a surgical team at New York University Langone Health (NYU). In the same month, a team at the Heersink School of Medicine, University of Alabama at Birmingham xenotransplanted a pair of the "ten-GE" pig kidneys into a brain-dead human recipient without hyperacute rejection before termination of the procedure after 74 hours. Limited kidney functions were recorded but it was considered that the major barriers to human xenotransplantation had been overcome.

I have very little envy but all the respect toward my immunologist colleagues, who had to work with very complex genetic, molecular, and cellular systems that interact in dynamic ways. It is not an overstatement that textbooks in immunology describe what we think we know how genes and immunity work together. It is an understatement that every human being has a different and unique edition of a personalized immunology textbook (**Figure 9-7**). Xenotransplantation still has a long journey ahead.

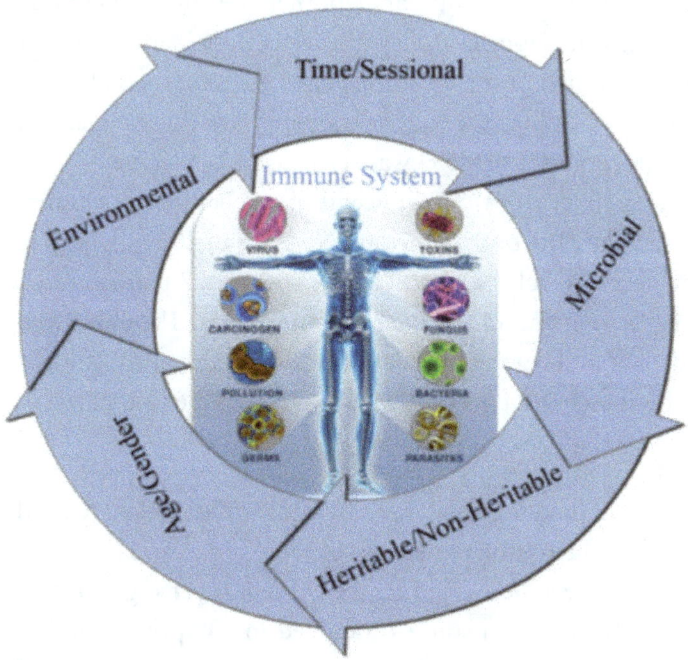

FIGURE 9-7. Immunity is personalized, too.

9.5 The Functions of Most Human Proteins Remain Largely Unknown

I am privileged to have had this opportunity to share with the readers my experiences and my views on the topics in human genetics. It would be fitting to end with closing remarks on how far we have come along and what remains to be pursued by the current and younger generations of scientists, and on what the general public can look forward to in the future.

The structure–function relationship of proteins, especially the variants resulting from a genetic change, is a very important topic in modern medicine. A 2016 survey of the molecular targets of 1,591 FDA-approved drugs showed that 1,194 (75%) act on 667 human-genome-derived proteins for various human diseases. This would represent about 667/20,000 = ~3.3% of the estimated total number of protein-coding genes in the human genome. The efforts spent and papers published on the human protein-coding genes appeared very disproportional. From 1980 to 2019, 596,891 papers were annotated to 20,122 human protein-coding genes. Among these, the single protein/gene p53, a tumor suppressor, had 9,243, about 1.5%, that is, there were three papers out of every 200 on one single protein/gene. In December 1993, p53 was selected as "Molecule of The Year" by *Science*, most certainly outshining the publicity of the EGFR with just a cameo appearance on the "*ER*" (**Chapter 5**). The EGFR was actually the bronze medalist (we are in 2024), with 5,287 papers, behind the silver medalist tumor necrosis factor alpha (TNFα), with 5,627 papers. At the same time, 623 human protein-coding genes were found to have zero publication. The publication numbers decreased sharply when plotted against the gene numbers (**Figure 9-8**), yet most had been cataloged in the Online Mendelian Inheritance in Man (OMIM, the blue band of a binary code 0 or 1) or included in GWAS.

In **Chapter 2**, we briefly outlined the challenges in "forward genetics" toward studying and understanding the functions of proteins, the regulation of their genes, and the variations among humans. Some challenges cannot be overcome, we then relied on "reverse genetics," using the approaches of "linkage analysis" and "association studies" detailed in **Chapters 7** and **8**. All efforts made not only left more than 600 human protein-coding genes behind, but the publication plot in **Figure 9-8** also looks very flat beyond 4,000 genes, that is, 80% of the total remain in what is called the "dark genome."

On June 26, 2000, the completion of the human genome map (first draft) was announced at a ceremony in the White House with the following statements by President Clinton:

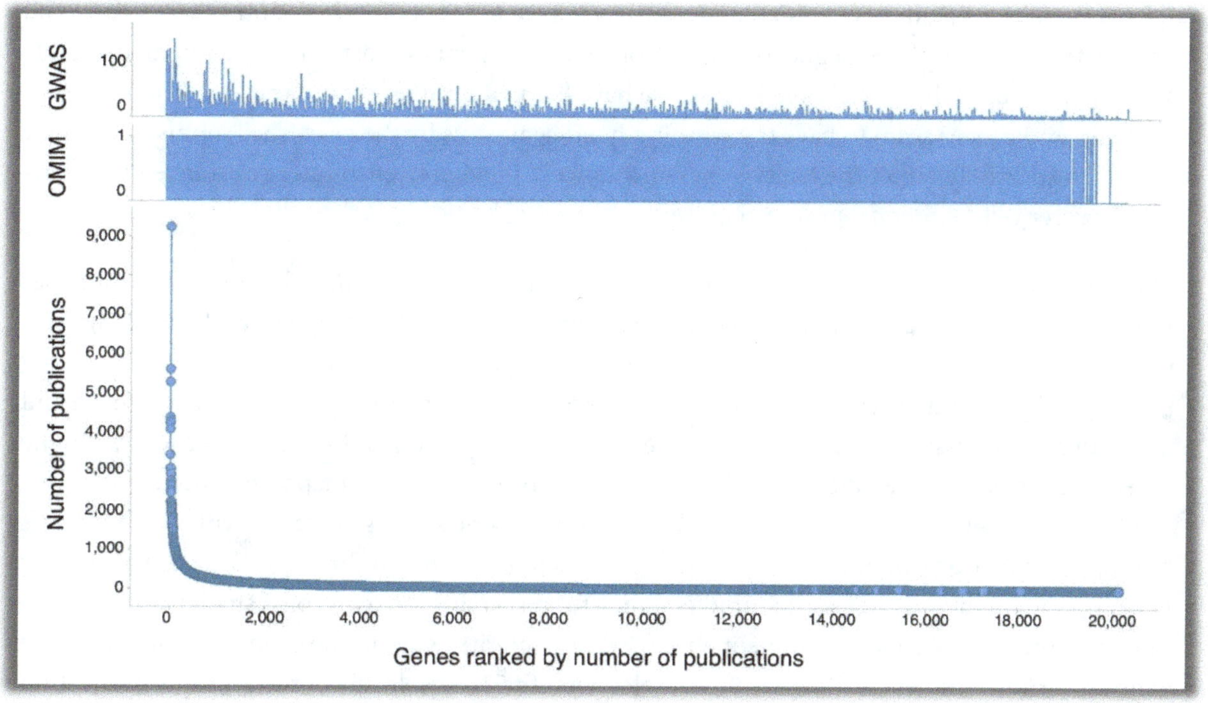

FIGURE 9-8. Publication numbers annotated to each human protein-coding gene

Today, we are learning the language in which God created life. We are gaining ever more awe for the complexity, the beauty, the wonder of God's most divine and sacred gift. With this profound new knowledge, humankind is on the verge of gaining immense new power to heal.

Genome science will have a real impact on all our lives and even more on the lives of our children. It will revolutionize the diagnosis, prevention, and treatment of most, if not all, human diseases.

In coming years, doctors increasingly will be able to cure diseases like Alzheimer's, Parkinson's, diabetes, and cancer by attacking their genetic roots. Just to offer one example, patients with some forms of leukemia and breast cancer are already being treated in clinical trials with sophisticated new drugs that precisely target the faulty genes and cancer.

It has been almost a quarter century, and clearly we have not accomplished all that was hoped and anticipated, especially "the diagnosis, prevention, and treatment of most, if not all, human diseases." We have not cured the Alzheimer Disease, we have learned even less about the Parkinson Disease. Diabetes and cancer seem to have the winning hand against us by the year. The "sophisticated new drugs that precisely target the faulty genes and cancer" inevitably encounter resistance by the targeted cancer cells, either *de novo* or acquired.

There are technical reasons for our lagging behind in the pace of fulfilling the hopes and anticipations, spoken by President Clinton toward the general public on the completion of such a massive publicly funded human research project. On the forward genetics end, identifying a disease-causing protein and purifying it remained difficult, for the lack of a proper assay to trace the presence and quantities of it in tissues and cells, or the protein exists in very low quantities and transiently. On the reverse genetics end, the parametric linkage analysis for disease gene mapping is hindered by the fashion and results of human mating, while the genome-wide studies are laden with false positives that take tremendous amounts of time and resources to sort out and through. There are also other reasons, one was described as a "natural tendency for researchers to focus on a small fraction of the human proteome." We had a comment about "whose favorite gene" in the discussion of "candidate gene association studies" in **Chapter 8**. The selection of p53 as molecule of the year was a reflection and also a proof of this tendency: "The excitement generated by it and its fellow tumor suppressors is reaching a crescendo with exhilarating possibilities for prevention and cure of cancer." As of February 2023, nearly 30 years later, very few drug development programs targeting p53 reached late-stage clinical trials, and no p53-targeting drug was approved in the United States or Europe.

Where do we go from here, and what should we do differently? In **Chapter 5**, we discussed the Beadle and Tatum "one gene, one enzyme" hypothesis proposed in 1941, most importantly the approach taken by them to probe for the functions of proteins by mutating the genes in a single cellular organism, the bread mold *Neurospora crassa* or *sitophila*. This concept has morphed into the framework of an ambitious initiative known as "Target 2035: Towards Medicines for All" by The EUbOPEN consortium: Enabling & Unlocking Biology in the OPEN, involving 22 academic and industrial partners. The approach is to develop "high-quality tools that would enable researchers to study a new protein." These include cell lines lacking the encoding gene, monoclonal antibodies, purified proteins with structural information, and pharmacological modulators for functional studies and the evaluation of a protein as a potential therapeutic target. The goal is to develop a freely available pharmacological modulator for every protein in the human proteome. On behalf of all geneticists, biologists, clinical scientists, and more important, the general public, we look forward to the exciting new findings with anticipation: "让我们拭目以待."

BIBLIOGRAPHY

1. https://www.nobelprize.org/prizes/chemistry/1993/mullis/facts/
2. https://www.genome.gov/about-genomics/fact-sheets/DNA-Sequencing-Costs-Data
3. https://www.organdonor.gov/learn/organ-donation-statistics
4. https://www.target2035.net/
5. Abourehab MAS, Alqahtani AM, Youssif BGM, Gouda AM. Globally approved EGFR inhibitors: Insights into their syntheses, target kinases, biological activities, receptor interactions, and metabolism. *Molecules.* 2021 Nov 4;26(21):6677.
6. Abramyan S, Hanlon M. Kidney transplantation. [Updated 2023 Jan 2]. In: StatPearls [Internet]. Treasure Island, FL: StatPearls Publishing; 2024 Jan. Available from: https://www.ncbi.nlm.nih.gov/books/NBK567755/
7. Anand RP, Layer JV, Heja D, Hirose T, et al. Design and testing of a humanized porcine donor for xenotransplantation. *Nature.* 2023 Oct;622(7982):393–401.
8. Armstrong JA. Urinalysis in Western culture: A brief history. *Kidney International.* 2007 Mar;71(5):384–7.
9. Blaese RM, Culver KW, Miller AD, Carter CS, et al. T lymphocyte-directed gene therapy for ADA-SCID: Initial trial results after 4 years. *Science.* 1995 Oct 20;270(5235):475–80.
10. Bock C, Datlinger P, Chardon F, Coelho MA, et al. High-content CRISPR screening. *Nature Reviews Methods Primers.* 2022;2(1):9.
11. Bubp J, Jen M, Matuszewski K. Caring for Glucose-6-phosphate dehydrogenase (G6PD)–deficient patients: Implications for pharmacy. Pharmacology & Therapeutics. 2015 Sep;40(9):572–4.
12. Carter AJ, Kraemer O, Zwick M, Mueller-Fahrnow A, Arrowsmith CH, Edwards AM. Target 2035: Probing the human proteome. *Drug Discovery Today.* 2019 Nov;24(11):2111–15.
13. Conner BJ, Reyes AA, Morin C, Itakura K, Teplitz RL, Wallace RB. Detection of sickle cell beta S-globin allele by hybridization with synthetic oligonucleotides. *Proceedings of the National Academy of Sciences of the United States of America* 1983 Jan;80(1):278–82.
14. Cooper DKC. Genetically engineered pig kidney transplantation in a brain-dead human subject. *Xenotransplantation.* 2021 Nov;28(6):e12718.
15. Cortés-Martín J, Sánchez-García JC, Rodríguez-Blanque R. Health care on rare diseases. *International Journal of Environmental Research and Public Health.* 2022 Dec 27;20(1):395.
16. Cruz-Tapias P, Castiblanco J, Anaya JM. Major histocompatibility complex: Antigen processing and presentation. In: Anaya JM, Shoenfeld Y, Rojas-Villarraga A, et al., editors. *Autoimmunity: From Bench to Bedside* [Internet]. Bogota, Colombia: El Rosario University Press; 2013 Jul 18. Chapter 10. Available from: https://www.ncbi.nlm.nih.gov/books/NBK459467/
17. Duan D. Dystrophin gene replacement and gene repair therapy for Duchenne muscular dystrophy in 2016: An interview. *Human Gene Therapy Clinical Development.* 2016 Mar;27(1):9–18.

18. Duan D, Goemans N, Takeda S, Mercuri E, Aartsma-Rus A. Duchenne muscular dystrophy. *Nature Reviews Disease Primers.* 2021 Feb 18;7(1):13.
19. Elangkovan N, Dickson G. Gene therapy for Duchenne muscular dystrophy. *Journal of Neuromuscular Distrophy.* 2021;8(s2):S303–16.
20. England SB, Nicholson LV, Johnson MA, Forrest SM, et al. Very mild muscular dystrophy associated with the deletion of 46% of dystrophin. *Nature.* 1990 Jan 11;343(6254):180–2.
21. Erkut E, Yokota T. CRISPR therapeutics for Duchenne muscular dystrophy. *International Journal of Molecular Science.* 2022 Feb 6;23(3):1832.
22. Fisman D. Cystic fibrosis heterozygosity: Carrier state or haploinsufficiency? *Proceedings of the National Academy of Sciences of the United States of America.* 2020 Feb 11;117(6):2740–2.
23. Garg AD. The dynamic interface of genetics and immunity: Toward future horizons in health & disease. *Genes & Immunity.* 2023 Aug;24(4):155–8.
24. Gazzoli I, Pulyakhina I, Verwey NE, Ariyurek Y, Laros JF, 't Hoen PA, Aartsma-Rus A. Non-sequential and multi-step splicing of the dystrophin transcript. *RNA Biology.* 2016;13(3):290–305.
25. Hamad E, Feldman AM. Pharmacogenetics in heart failure: How it will shape the future. *Heart Failure Clinic.* 2010 Jan;6(1):1–10.
26. Hassin O, Oren M. Drugging p53 in cancer: One protein, many targets. *Nature Reviews Drug Discovery.* 2023 Feb;22(2):127–44.
27. Hildyard JCW, Piercy RJ. When size really matters: The eccentricities of dystrophin transcription and the hazards of quantifying mRNA from very long genes. *Biomedicines.* 2023 Jul 24;11(7):2082.
28. InformedHealth.org [Internet]. Cologne, Germany: Institute for Quality and Efficiency in Health Care (IQWiG); 2006. In brief: Understanding urine tests. [Updated 2023 Feb 27]. Available from: https://www.ncbi.nlm.nih.gov/books/NBK279350/
29. Iqbal N, Iqbal N. Imatinib: A breakthrough of targeted therapy in cancer. *Chemotherapy Research and Practice.* 2014;2014:357027.
30. Koenig M, Hoffman EP, Bertelson CJ, Monaco AP, Feener C, Kunkel LM. Complete cloning of the Duchenne muscular dystrophy (DMD) cDNA and preliminary genomic organization of the DMD gene in normal and affected individuals. *Cell.* 1987 Jul 31;50(3):509–17.
31. Koshland DE Jr. Molecule of the year. *Science.* 1993 Dec 24;262(5142):1953.
32. Kulski JK, Suzuki S, Shiina T. Human leukocyte antigen super-locus: Nexus of genomic supergenes, SNPs, indels, transcripts, and haplotypes. *Human Genome Variation.* 2022; 9:49.
33. Kunkel LM, Monaco AP, Middlesworth W, Ochs HD, Latt SA. Specific cloning of DNA fragments absent from the DNA of a male patient with an X chromosome deletion. *Proceedings of the National Academy of Sciences of the United States of America.* 1985 Jul;82(14):4778–82.
34. Kunkel LM. 2004 William Allan Award address. Cloning of the DMD gene. *American Journal of Human Genetics.* 2005 Feb;76(2):205–14.

35. Li Q, Lan P. Activation of immune signals during organ transplantation. *Signal Transduction and Targeted Therapy.* 2023; 8:110.
36. Maiuri L, Raia V, Kroemer G. Strategies for the etiological therapy of cystic fibrosis. *Cell Death & Differentiation.* 2017 Nov;24(11):1825–44.
37. Mall MA. ENaC inhibition in cystic fibrosis: Potential role in the new era of CFTR modulator therapies. *European Respiratory Journal.* 2020 Dec 24;56(6):2000946.
38. Mallapaty S, Kozlov M. First pig kidney transplant in a person: What it means for the future. *Nature.* 2024 Apr;628(8006):13–14.
39. Middleton PG, Mall MA, Dřevínek P, Lands LC, et al., VX17-445-102 Study Group. Elexacaftor-Tezacaftor-Ivacaftor for cystic fibrosis with a single Phe508del allele. *New England Journal of Medicine.* 2019 Nov 7;381(19):1809–1819.
40. Moini J, Logalbo A, Schnellmann JG. Chapter 5—Pharmacogenomics, drug toxicity, and environmental toxins, Editor(s): Moini, Logalbo, Schnellmann, *Neuropsychopharmacology*, Academic Press, 2023, pp. 75–96.
41. Monaco AP, Neve RL, Colletti-Feener C, Bertelson CJ, Kurnit DM, Kunkel LM. Isolation of candidate cDNAs for portions of the Duchenne muscular dystrophy gene. *Nature.* 1986 Oct 16–22;323(6089):646–50.
42. Neri M, Rossi R, Trabanelli C, Mauro A, et al. The genetic landscape of dystrophin mutations in Italy: A nationwide study. *Frontiers in Genetics.* 2020 Mar 3;11:131.
43. Olson EN. Toward the correction of muscular dystrophy by gene editing. *Proceedings of the National Academy of Sciences of the United States of America.* 2021 Jun 1;118(22):e2004840117.
44. Pirmohamed M. Pharmacogenetics and pharmacogenomics. *British Journal of Clinical Pharmacology.* 2001 Oct;52(4):345–7.
45. Pirmohamed M. Pharmacogenomics: Current status and future perspectives. *Nature Reviews Genetics.* 2023 Jun;24(6):350–62.
46. Porrett PM, Orandi BJ, Kumar V, Houp J, et al. First clinical-grade porcine kidney xenotransplant using a human decedent model. *American Journal of Transplantation.* 2022 Apr;22(4):1037–53.
47. Pranke I, Golec A, Hinzpeter A, Edelman A, Sermet-Gaudelus I. Emerging therapeutic approaches for cystic fibrosis. From gene editing to personalized medicine. *Frontiers in Pharmacology.* 2019 Feb 27;10:121.
48. Rana A, Godfrey EL. Outcomes in solid-organ transplantation: Success and stagnation. *Texas Heart Institute Journal.* 2019 Feb 1;46(1):75–6.
49. Reddy EP, Aggarwal AK. The ins and outs of bcr-abl inhibition. *Genes & Cancer.* 2012 May;3(5–6):447–54.
50. Reynolds EH, Broussolle E. Allbutt of Leeds and Duchenne de Boulogne: Newly discovered insights on Duchenne by a British neuropsychiatrist. *Revue Neurologique* (Paris). 2018 May;174(5):308–312.

51. Richardson SR, O'Malley GF. Glucose-6-phosphate dehydrogenase deficiency. [Updated 2022 Sep 26]. In: *StatPearls* [Internet]. Treasure Island, FL: StatPearls Publishing; 2024 Jan. Available from: https://www.ncbi.nlm.nih.gov/books/NBK470315/
52. Saiki RK, Scharf S, Faloona F, Mullis KB, Horn GT, Erlich HA, Arnheim N. Enzymatic amplification of beta-globin genomic sequences and restriction site analysis for diagnosis of sickle cell anemia. *Science.* 1985 Dec 20;230(4732):1350–4.
53. Santos R, Ursu O, Gaulton A, Bento AP, et al. A comprehensive map of molecular drug targets. *Nature Reviews Drug Discovery.* 2017 Jan;16(1):19–34.
54. Schwartzberg L, Kim ES, Liu D, Schrag D. Precision oncology: Who, how, what, when, and when not? *American Society of Clinical Oncology Educational Book.* 2017;37:160–9.
55. Sharabi AF, Goudar RB. Alkaptonuria. [Updated 2023 Aug 8]. In: *StatPearls* [Internet]. Treasure Island, FL: StatPearls Publishing; 2024 Jan. Available from: https://www.ncbi.nlm.nih.gov/books/NBK560571/
56. Shchaslyvyi AY, Antonenko SV, Tesliuk MG, Telegeev GD. Current state of human gene therapy: Approved products and vectors. *Pharmaceuticals* (Basel). 2023 Oct 5;16(10):1416.
57. Shiferaw D, Faruqi S. Profile of tezacaftor/ivacaftor combination and its potential in the treatment of cystic fibrosis. *Therapeutics and Clinical Risk Management.* 2019 Aug 22;15:1029–1040. doi:10.2147/TCRM.S165027. Erratum in: *Therapeutics and Clincal Risk Management.* 2019 Oct 11;15:1207.
58. Shiravand Y, Khodadadi F, Kashani SMA, Hosseini-Fard SR, Hosseini S, Sadeghirad H, Ladwa R, O'Byrne K, Kulasinghe A. Immune checkpoint inhibitors in cancer therapy. *Current Oncology.* 2022 Apr 24;29(5):3044–3060.
59. Starner CI, Gleason PP. Spinal muscular atrophy therapies: ICER grounds the price to value conversation in facts. *Journal of Managed Care & Specialty Pharmacy.* 2019 Dec;25(12):1306–8.
60. Tewkesbury DH, Robey RC, Barry PJ. Progress in precision medicine in cystic fibrosis: A focus on CFTR modulator therapy. *Breathe* (Sheff). 2021 Dec;17(4):210112.
61. Tsongalis GJ, Silverman LM. Molecular diagnostics: A historical perspective. *Clinica Chimica Acta.* 2006 Jul 23;369(2):188–92.
62. Venugopal V, Pavlakis S. Duchenne muscular dystrophy. [Updated 2023 Jul 10]. In: *StatPearls* [Internet]. Treasure Island, FL: StatPearls Publishing; 2024 Jan. Available from: https://www.ncbi.nlm.nih.gov/books/NBK482346/
63. Wen X, Luo Z, Zhao W, Calandrelli R, Nguyen TC, Wan X, Charles Richard JL, Zhong S. Single-cell multiplex chromatin and RNA interactions in ageing human brain. *Nature.* 2024 Apr;628(8008):648–56.
64. Wu DY, Ugozzoli L, Pal BK, Wallace RB. Allele-specific enzymatic amplification of beta-globin genomic DNA for diagnosis of sickle cell anemia. *Proceedings of the National Academy of Sciences of the United States of America* 1989 Apr;86(8):2757–60.

Postface

Making “sense” of genetics today can be overwhelming, especially with the innovations in technology, which brings deeper and broader insights into the biological world. It is the smallest thing but keeping up with the acronyms of all the next-gen multi-omics is already not easy. “Next-gen” and “omics” have the risks of soon being replaced, too. For myself, as a university professor and researcher, the sense I want to make from reading the reports today on the advances in genetics is “SO WHAT.”

“SO WHAT” was a required element in the annual progress and final reports to the Department of Defense Prostate/Breast Cancer Research Programs (DoD-PCRP and DoD-BCRP), part of the Congressionally Directed Medical Research Programs (CDMRP). These programs invited “consumer reviewers” on the review panels of research proposals submitted to the programs for funding. The consumer reviewers were cancer survivors and advocates who served as voting members to score on the merits of the proposed research. In the research proposal, the applicant scientists were also required to provide a "Public Abstract" along with the "Technical Abstract." In my years of active research funded by the DoD cancer research programs and service on the review panels, I fully embraced the idea of the "SO WHAT" (all caps) in the reports and the public abstract in the proposal.

Making sense of genetics and getting out the "SO WHAT" from the reports of today's news in genetics are part of every family's everyday life. Over the years, thanks to my mother being proud of what I do in my professional life, I have talked to many relatives, and her former coworkers and even her college friends from the 1950s. They were all physicists and engineers, with little or no background in biology, and specifically and significantly less in genetics or molecular genetics. The focus was mostly on questions regarding cancer treatment and survival, seeking information for themselves or their loved ones. One common theme in the conversations was "targeted therapy," or more specifically, "molecular targeting therapy" of cancer. The folks I spoke to were all in China, where molecular targeting therapy can be very expensive, such as the monoclonal antibody trastuzumab (Herceptin) for Her2+ breast cancer, used by a first cousin of my mother's before the cousin passed away a few years ago. I found myself completely unprepared in the beginning but getting better with time on how to explain what a molecular target in cancer meant and why these targets do not necessarily mean hitting it will kill off the cancer. Cancers will show resistance to drugs, targeted or not, *de novo* or acquired. To make such points clear, a proper crash course in gene expression regulation and chromosome (chromatin) structures was always needed. Of course in the end, the decisions on the treatment options were always made by the patients in consultation with their physicians.

My mother at age 85 is now a 27-year survivor of a T2a bladder cancer that required a partial cystectomy in the summer of 1997, and the multiple recurrences over the years. I do not provide her with medical care, but she found it comforting to have someone immediately available to help understand the meaning of every technical term in her medical records. In May of 2023, I was given the honor and privilege to ask the first technical question to my older son, Andreas, after his dissertation research presentation at his defense, by his mentor, Dr. Eric Olson. One chapter of Andreas's PhD dissertation work was a paper published in *Nature Medicine* in February of 2023, titled "Base Editing Correction of Hypertrophic Cardiomyopathy in Human Cardiomyocytes and Humanized Mice," which deservedly caught a good amount of media attention, too. This is the most common form of genetic heart disease. Variants in cardiac sarcomeric genes cause abnormal heart muscle thickening, increasing the risks of heart failure, arrhythmia, and sudden cardiac death. About 20%–30% of familial hypertrophic cardiomyopathy (HCM) cases had mutations in the gene encoding β-myosin heavy chain 7 (MYH7). Andreas used base editing, a form of gene editing, to correct the dominant-negative c.1208G>A (p.R403Q) pathogenic MYH7 variant, causal to increased cardiac contractility and HCM. The efforts were successful in human cardiomyocytes and in mice carrying the exact human variant with similar pathology.

I asked Andreas (unrehearsed): "What is the mode of inheritance and what is the age of onset for HCM?"

Andreas: "For the specific pathogenic variant it is autosomal dominant, and the age of onset of HCM is between 20 to 50 years of age."

I replied: "Of course we don't have the genotype information, but your grandfather, whom you did not get to meet, died at the age of 49 from sudden cardiac death. His heart was shown by autopsy to be twice the normal size. His father died at the age of 61 from a similar event. I am 58, and if I make it to 62, you and your brother should be fine."

That was an easy example of making sense of genetics and I was speaking to an expert about to earn a PhD (he did!). There is quite a lot that I cannot make much sense of. Everything is investigated on a single-cell level, now. We have seen an example of looking for genetic clues of Alzheimer Disease in single brain cells. The journal *Genes & Immunity* was "revamped" in August of 2023, toward "single cell or spatial -omics, and high-end imaging" and "deep profiling of different immune cell subsets" for "immunogenomic, immuno-transcriptomic, or immuno-proteomic big data," as well as "computational immunology and the usage of artificial intelligence."

The same has been done in cancer research for many years, now. At the beginning of the cancer research single-cell movement, I attended a seminar given by a researcher from a renowned cancer center in the United States. He presented his findings of single-cell sequencing of the genomic DNA from 1,000 different cells, 500 from one single breast cancer nodule and 500 from another nodule, both from one breast of a cancer patient. The main point of the findings: all 1,000 cells are genetically different. The intended utility of the information: precision medicine. I will defer making sense of this to artificial intelligence.

www.ingramcontent.com/pod-product-compliance
Lightning Source LLC
LaVergne TN
LVHW080049210926
848481LV00011B/3314
9798385111411